Energy in America

Fifteen Views

The **Center for Study of the American Experience** seeks to trace the evolutionary development and examine the future prospects of selected threads drawn from the unique fabric of American life, and to perform analysis and engage in creative thought on problems and opportunities that have been most influential in shaping America's past and present and will be most likely to define America's future.

The objectives are pursued through the assembling of conferences at the highest level of objectivity and expertise, the production of edited audio and visual tapes and publications from these conferences, and the support of visiting distinguished Scholars-in-Residence.

The Center is a joint educational project of the University of Southern California and The Annenberg School of Communications located on that campus. It welcomes the intellectual involvement of the University's faculty and students, as well as leaders of thought from the nation at large.

Topics for conferences are selected to stimulate awareness of the various aspects of the American experience, the potential benefits of future opportunities, and the possible solutions to existing and emerging problems of America.

Simon Ramo
Chairman

John C. Weaver
Executive Director

Other titles in the Center for Study of the American Experience series

Rights and Responsibilities:
International, Social, and Individual Dimensions

The Arts in Shaping the American Experience

**Proceedings of a Conference
Sponsored by the
Center for Study of the American Experience
The Annenberg School of Communications
University of Southern California**

May 1979

Participants

Simon Ramo, *Conference Chairman*

Arthur M. Bueche
W. Kenneth Davis
Selwyn Enzer
John S. Foster, Jr.
Charles J. Hitch
William W. Hogan
John P. Holdren
J. Clayburn La Force
Henry R. Linden
Bruce Murray
Dale D. Myers
Sam H. Schurr
David Sternlight
John E. Swearingen
Jerome B. Wiesner

Energy in America

Fifteen Views

University of Southern California Press
Transaction Books • Rutgers University

Published 1980
University of Southern California Press
Los Angeles, California 90007

Series Editor, Joyce J. Bartell
Designer, Robert W. Giese

The "Fifteen Views" on various facets of America's energy picture recorded in this volume were supplemented, reinforced, or challenged in questions and discussion by the audience. The extemporized remarks were electronically taped, then transcribed and edited for clarity and brevity. While not always verbatim as reported here, they convey the tenor of response given each paper delivered, and multiply fifteen views to several times fifteen. It was not possible to identify all speakers.

J.J.B.

Distributed by Transaction Books Rutgers University
New Brunswick, N.J. 08903 (U.S.A.) and London (U.K.)

Library of Congress Cataloging in Publication Data
Main entry under title:
Energy in America.
1. Power resources—United States—Congresses.
2. Energy policy—United States—Congresses.
I. Ramo, Simon. II. Annenberg School of Communications, University of Southern California. Center for Study of the American Experience.
TJ163.15.E534 333.79'0973 80-36739

ISBN 0-88474-103-6

CONTENTS

Simon Ramo

Introduction

Fifteen nationally recognized authorities in the technology and economics of energy met at the Center for Study of the American Experience in May 1979 to ponder an issue many believe soon will emerge as the core of the U.S. energy problem—the vital tradeoff between the supply and the price of energy in the future.

Everyone expects the price of energy to rise—but will it go up and up, or level off at a new plateau? At some price can we have a copious supply? Can the United States enjoy economic prosperity at a higher price level if the energy supply is plentiful? What will happen to employment, economic growth, inflation, the dollar?

The energy experts were from academia, industry and government. Addressing the topic of "America in an Era of Plentiful Although Higher Priced Energy," each gave his views and then was questioned by an invited audience of a hundred energy specialists. They came and left with differences of opinion. Yet the areas of agreement were broad and portend both optimistic and pessimistic futures.

To oversimplify slightly, they concluded that at a price about double that being paid for crude oil at the time of the conference, the supply of available energy would more than equal the demand. At such a price many options become economically sound:

Simon Ramo, *Chairman of the Center for Study of the American Experience, is the "R" and a co-founder of TRW Inc. Recently honored with the National Medal of Science, the nation's highest science award, he has been chief scientist of the Intercontinental Ballistic Missile Program and Chairman of the President's Committee on Science and Technology. He is a Fellow of the National Academy of Sciences, a Trustee and Visiting Professor in Management Sciences at the California Institute of Technology, and Fellow of the Faculty of the Kennedy School of Government at Harvard University.*

Dr. Ramo served as Chairman of the energy conference.

- For conventional petroleum sources, the higher price would cover more expensive exploration and extraction (as from the ocean wells offshore, deeper land drilling, and forced recovery from existing wells).
- Oil from shale and tar sands would become practical.
- Coal could be mined more safely, desulfurized before burning, burned more cleanly, and energy extracted from it without mining (in situ).
- A synthetic gas and liquid fuel industry would yield adequate return on the necessarily huge investment.
- Nuclear reactors could be built and located so as to operate with greatly added safety.
- Solar installations would make a noticeable contribution to the total supply.
- More natural gas and gas from unconventional sources could be obtained.

In short, at this higher price, the economics and technology of the supply problem would harmonize, and this match-up would curb further price rises.

The experts also shared the conclusion that the American economy could accommodate to this higher price level. Of course, specific industries and segments of the population would experience more problems of adjustment than would others. But the average price structure for all activities of the nation would rise only modestly in response to a doubling of the basic cost of raw energy resources. An economy not handicapped by energy shortages would automatically be fundamentally stronger and the general strength would compensate for the energy price rise and permit aid to adversely affected individuals.

In contrast, a big cut in supply, whatever the price, would so dislocate manufacturing, transportation and distribution as to cripple the economy. Even at a high price, a plentiful domestic supply means the funds expended stay within the nation instead of being exported to buy oil.

The gathered savants realized these conclusions raise a basic question. If new discoveries and developments can provide ample energy at a price that, although higher, is economically containable, then why the present difficulties? Why the concern about the U.S. energy future? The reasons, the conference attendees believe, lie not in technology or economics but in social and political confusion.

Examples? There are many. For instance, government policies have preached conservation while subsidizing lower prices, thus removing incentives to conserve. A higher price for automobile fuel would have shifted the public more rapidly to conservation. Also, one part of the government mandated a shift-over of utilities from oil to coal while another part placed such stringent environmental controls on the use of coal as to make the shift virtually impossible. An adjusted, gradual compromise of these conflicting requirements has been too difficult to arrange.

Also, the experts tell us the reserves of oil available domestically could be doubled if advanced technology for recovery were used on all appropriate existing wells but such procedures are expensive and thus practical only with a higher price structure than the government has been ready to allow.

Still another example: at true market prices for energy, private investment would have accelerated development of energy alternatives and government funds could have been concentrated on building badly needed, improved urban mass transit.

Social-political progress, the experts concluded, is needed to make the transition to a new era of plentiful, though higher priced, energy. In the United States, the handling of the energy problem absolutely requires a partnership of government and the private sector, with both government and the private sector determined to team up instead of continuing the present, harmful adversary relationship.

The government cannot do it by itself because it lacks the overall competence and is controlled by political considerations pitting constituent groups against one another. Private industry alone cannot give us the energy we need because the enormous interaction problems—transportation, environment, safety hazards, land use, water allocation, natural resource conservation, to name only a few—require national policies and substantial regulation.

Recognizing how difficult it is in America to bring the necessary parties together without a "Pearl Harbor" to cause the national interest to dominate, some of the experts said things must get worse before they can get better.

Our difficulty is with the transition period we are in right now. How long this unsatisfactory period lasts depends upon the nation's speed in social-political maturing, and not on technology or economics. Ahead there can be an era of economic prosperity for the United States based on a plentiful, although higher-priced, supply of energy.

Jerome B. Wiesner

Can Society Survive the Energy Crunch?

Several years ago, the President challenged the nation to treat the problem of energy resources as "the moral equivalent of war." So far the people of America have not been able to respond with so much as a good pillow fight. Maybe this conference can shed some light on why we remain so confused.

I plan to talk about the energy situation from two points of view, on neither of which I am a professional. I shall talk about the specifics of the energy situation and what I might do if I were in charge. And I also want to explore a basic question: why a vital task like exploiting available energy resources seems to be so difficult in the United States today, and why even conservation, in spite of its importance, is so hard to get started. Or to put it more bluntly, why we have apparently shifted from a nation of doers to a nation of complainers that cannot seem "to get its act together."

Obviously, if we are correct when we say that the United States has ample energy resources, then something must be inhibiting exploitation of them. Something is. And I think it stems from society's growing inability to respond to new circumstances and new problems, to reconcile its conflicting demands between growth and conservation, and

---

Jerome B. Wiesner *has been on the faculty of the Massachusetts Institute of Technology since 1946 as Professor, Director of the Research Laboratory of Electronics, Dean of Science, Provost, and (since 1971) President. He was special assistant to President Kennedy on science and technology and Chairman of his Science Advisory Committee. He has also headed the advisory committee to the Office of Technical Assessment of the United States Congress.*

particularly, its total confusion regarding its actual options when the boundary conditions such as resources, environmental destruction, risk, and so on, are factored in—all important items about which all citizens are concerned and do not know how to bring into balance. So I want especially to relate the energy quandary to these deeper issues, and my goal is to try to convince you of my view of how a democratic, industrial, capitalistic society functions, and how such insights will help us all to understand the energy problems and, it is hoped, to respond more effectively to them.

It is obvious to me that industrial societies all over the world are having to adjust to a wholely new situation caused by an increasing scale, sophistication, interconnectedness, and increasingly dynamic character of their societal systems. Their response has been primarily regulations and controls and governmentally-managed projects. Because the functioning of a complex industrial system is poorly understood even by economists, the interventions often as not are counterproductive. And even those interventions that are clearly essential and desirable—such as environmental safeguards, or eliminating risks from drugs or food additives, or increasing product reliability—have costs both in time and money that need to be better understood and factored in the decision-making processes before they are undertaken.

Society as a Learning Machine

As a communications engineer I have had a life-long interest in human and man-made communication and learning systems, the area that Norbert Wiener dubbed "cybernetics." I have tried to understand our society as a learning machine model, a system continually required to solve new problems. It will come as no surprise to any of you that my conclusion is that our society does not come off very well when so judged; however, it is in fact learning (if too slowly), and it would not be hard to make it do much better—something that ought to be its goal.

Because of my experience, first as Presidential Science Advisor and now as President of M.I.T., the media and many other groups turn to me when they seek explanations of specific technical events such as the Three Mile Island accident, or when they are looking for a whipping boy, and particularly when they want a spokesperson identified with science and technology. I seem to be a special target whenever the social roles of science and technology are being examined critically. Through the years, this exposed position has given me a unique opportunity to hear the concerns of the general public and the many special interest groups that vehemently oppose specific technology—such as nuclear power, or technical and industrial growth—and that has forced me to try to understand their laments in order to contend with them.

Not long ago I was asked to be the "scientist" on a panel at an annual meeting of the Explorers Club. The subject for the evening was "Survival," and I thought they meant "survival" in the old-fashioned sense and wanted to hear about a boating accident I had had a few years ago. But I discovered they had a much more serious intent: they were concerned about the survival of humanity. I speculated then, just as I speculate now, on what the question means. What does "survival" mean? And I tried to find out what made scholars like Paul Ehrlich and economist Robert Heilbroner and the author of "Small Is Beautiful," Schumacher, so pessimistic and doubtful that the kind of world most of us are fond of can possibly survive.

In reading and in talking with some of these people I found that basically they did not believe that a free society based on technology, supported by technology, could cope with problems as fast as they emerged; that a combination of resource shortages, inevitable population growth, inevitable environmental damage from such things as carbon dioxide in the atmosphere, technological failures, and a general inability to manage our affairs adequately, would so discourage the general citizenry that the free enterprise system would collapse. These views, I find, are shared by millions of people around the world as well as here in the United States, even while our system provides such unprecedented well-being. Furthermore, it is the continuing emergence of new problems that is most discouraging to the average person.

A serious consequence of this troubled state of mind and the fear and suspicion there is of technology is a growing mistrust of all our institutions of society—business, government, the university, the press—so that many, many people automatically discount what anyone from these sectors of the society says. I will argue that surprises such as those that trouble many people are, and always will be, a necessary feature of efforts to improve the society; and that they should be expected, detected, and corrected rather than avoided, because they can be avoided only by stopping the introduction of new technologies (which, in fact, is advocated by some), a course that clearly leads us downhill all the way. The challenge, it seems to me, is to get a widespread understanding of this nature of a modern and continuously evolving society.

It is interesting for me to note, by way of contrast, that essentially all leaders of the less advanced countries—China, Russia, India, Egypt, and many others—believe that a high level of technology is an essential element of their efforts to eliminate poverty, disease, and illiteracy in their countries. And it is also equally interesting to note that they and much of the rest of the world look to and depend upon us for the technology they need to succeed in their objectives.

Two Basic Problems

During my "survival" study, I concluded that our country faces only two basic problems, and they are the ones I focus on here: first, insuring an adequate supply of environmentally and economically compatible energy; and second, learning how to make the society more responsive to changing conditions—in other words, how to make it a better learning machine. The first, obviously, depends on the second.

Our current confusion and apparent inability to move effectively on these problems arises because a fundamental discontinuity has occurred in the human situation during the last half-century. Our vast man-made systems have completely altered the human relationship to the natural world and created a situation where almost everything affects everything else, thus calling for a continuing and rapid adaptation to changing natural conditions and to the man-made environment at the same time, a situation that I think is not generally appreciated.

This is really quite new, because for most of man's long history, the great challenge to survival came from the natural environment—that is, of acquiring enough food and providing shelter from the hostile environment in order to survive. For a very long time, progress in this struggle was slow and painful; but during the past century, organized science and technological industry so successfully multiplied the forces at man's disposal and so completely created a synthetic environment, that nature's threats have been pushed into the background, and some new ones, less familiar and therefore more worrisome than nature's—like nuclear power, environmental pollution, and many others—have come to the fore, at least for the moment. This change has not been adequately understood.

Also, I suspect that many people believed, even expected, that a technological utopia would emerge about now—or at least, sometime—in which life would be totally safe and predictable; and they grow increasingly upset by the new troubles and the fact that it is necessary to continue to struggle hard just to cope with surprises that keep cropping up.

I believe human existence will always be thus, so everyone had better get used to the idea. And one of the jobs of those in education, in government, in communications, I think, is to get people to appreciate this fact of life. This leads to my major point.

Modern societies can exist only in a state of dynamic equilibrium that involves continuing and often rapid adaptation to the changing man-made world, and to the natural world that changes because of man's activities. All societies and their manifold subsectors, such as the energy industry, are complex, multiply-interconnected, dynamic learning systems, whose effectiveness should be judged by how well they learn and adapt. And as one tinkers with them, the major criteria should be, is the ability of the systems to adapt and learn being improved? Much of the misunderstanding and alienation that exists arises because people do not realize the essential nature of the difficult and halting process by which individuals and societies learn, so they are prone to believe that most of their difficulties are the result of human malfeasance and could have been avoided.

"Feedback" in the Learning Process

Whether people realize it or not, and whether they like it or not, learning normally occurs through an experimental or trial-and-error process. That involves selecting a goal, taking a tentative step forward, and comparing the result with the objective. Professionals in the controls systems business call this "feedback." And in the feedback system, if the result appears to be in the right direction, one takes a second step; if the step appears to be in the wrong direction, corrective measures are called for. Societies must learn in the same way, and they must learn for two reasons: first, as I have already said, just to maintain their status quo as conditions change; and second, as one would like, to improve the conditions of life for all their citizens.

In the United States at the moment and for the last several decades, people have been working hard on both goals. We Americans are struggling to maintain our industrial and social advances in the face of resource limitations and the impact of large-scale technology on the physical environment; and we are simultaneously trying to improve the quality of life and the level of opportunities for everyone. There are often conflicts between these two sets of goals, which must be reconciled by the political process, and in the energy dilemma I think we are caught right in the middle of such a situation.

It is ironic that the more dynamic the learning process (and presumably, progress) is, the greater will be the uncertainties and the strains in the society. And because so few people understand that the trial-and-error process is essential to societal evolution, the more rapidly the problems are identified and an effort made to solve them, the more upset people are likely to be. How many people do you suppose think that Three Mile Island will be regarded as a major and positive learning experience for our nuclear industry and our nation? I happen to think it will when it is all shaken down.

The political system and social structure of a country obviously affect the quality and character of the societal learning process. In a truly free enterprise system most of the decisions regarding the allocation of resources happen automatically as a result of a complex summation of individual choices and gambles. This situation encourages experimentation, provides a simple measure of success or failure, and allows the system to function without constant intervention and realignment.

It is perfectly obvious, I think, that there are no longer any wholely free enterprise economies among the technologically-based nations. As societies have become more complex, there has been a great increase in the number of areas where individual initiatives generated problems that required a cooperative mode for a solution—that is, required the government to act for the whole, either to provide a needed service, such as education or defense, or to set bounds on individual initiatives in health and environmental actions. These actions, though frequently necessary, slow down the learning process, often seriously. Sometimes the delays are tolerable and perhaps even desirable; in most cases, however, their effect is quite deleterious. Those of you who have studied feedback control systems know what happens when the feedback loops of a dynamic system are improperly designed. If the feedback circuit has too many and too lengthy delays, the system will oscillate violently.

"Feedback" in an Economic System

I believe this effect can be seen in our economic system and in all economic systems, whether they are totally free or managed and manipulated. In this sense, the current tax revolt can be regarded as a feedback signal—somewhat late and weak, but nevertheless a feedback signal. If such a signal is dominated by noise (and maybe this one is), the system lashes about in a kind of random and wild fashion; that can sometimes be seen, too, in our society. If there is too much noise—that is, bad information in the system—it will also hunt rather aimlessly. Furthermore, if the sensitivity of the error-detecting element of the control system is too small or too low in an otherwise adequate feedback system, large errors will be required to activate corrective measures. These phenomena are all evident—too evident—in the societal learning system, and particularly in the public sector where feedback that comes from price competition is almost completely lacking, as in the consumer sector of the socialist countries.

Enough is known about the dynamics of physical systems to design them so that these pathological effects do not occur. I believe that enough is also known about society's problems to improve its competence to learn, though it is obviously not possible to do the same kind of analytical design that is done for machines. But just understanding what is going on in our man-made world should allow better management; knowing that each new initiative is an experiment that could cause new problems allows concentration on a search for solutions instead of a search for villains. It should also stimulate a search for solutions where automatic forces such as the markets rather than regulation dominate, within general bounds set by broad societal goals.

It is extremely hard to anticipate all the effects of changes, and therefore, many well-intentioned initiatives get into difficulty. The federal housing program of the '60s is a good example of this; I think the oil and gas price regulation is, too, and you can think of many others. The challenge then, is to seek ways to improve this societal learning system to create a society that can learn more effectively to detect and understand its emerging problems, and then to make many quick, intelligently-planned, and not too large experiments to find ways to correct them.

Replacement Technologies

Today's problems, the problems of this conference and the energy field, are made especially difficult by the growing need for what I call substitute, sophisticated, replacement technologies—that is, technologies to replace our existing and obsoleting systems as various factors in the society such as energy prices and environmental problems make them less useful. Creating and putting such replacement technologies in use is as a rule much more difficult than it was to create the original systems. The earlier forms were developed slowly and steadily in an evolutionary manner over a period of many years, growing in size and improving in reliability, but the new technologies not only are typically much more complex, larger, and more costly than those they replace, but they must meet more severe standards of efficiency, environmental impact, safety, liability, and so on.

It is in this area that the much-discussed innovation gap is most apparent. When one takes into account the amount of effort and resources that must go into these replacement technologies, and the fact that the societies have created and are creating elaborate regulatory safeguards against the feared side-effects of all new technologies (which generally increase the time required for testing, approval, and so on), there is a much greater danger today, I think, of producing violent and traumatic discontinuities in our society by the failure to have needed new capabilities when they are needed than by their impacts. And this is particularly true, I believe, in the energy area.

For needed sophisticated technologies to come into being, long-range undertakings are required that go beyond industry's normal investment time horizon, particularly in the mood created by the present regulatory uncertainty and the tax environment. To make viable some of the replacement technologies, such as synthetic fuel processes, business tax and regulatory innovations are at least as important as new technical developments. Perhaps the German government's practice of providing matching front-end money for risky long-term projects is the path to follow here. The key to success in such ventures will be, I believe, the degree to which the arrangements put the initiatives back into the hands of the individual entrepreneurial groups, and allow them to set their own pace, take their own risks, and compete with others like themselves.

As I worried about "survival," it quickly became apparent that much of the widespread confusion about energy issues stems from an inability to separate the immediate, the intermediate, and the long-term problems; and the economic, technical and societal problems, which are all mixed up together; and one further pair: technical issues must be divided into the scientific questions, which really limit our dreams, and the engineering problems of getting the possible done. So when somebody speaks of the energy problem, one has to ask what subset of these issues he has in mind. Unless these matters are considered separately and one by one, people will always be confused about our energy situation.

Turning for just a few minutes to the specific issues of this conference, given today's high price of crude oil on the world market one would expect that the market economy would stimulate our entrepreneurs in fields and laboratories to find ways to exploit the available resources. But actually, price control in the United States and environmental problems and other societal issues have up to now created more disincentives than incentives for new energy production. I hope that the

President's recent actions on deregulation will stick, and that the tax credits for shale oil will help correct the problem in that area. The new situation will certainly stimulate the search for oil, and probably encourage as well as enhance recovery techniques. However, it might take more than tax credits to create a gusher of shale and other synthetic fuels; these sources are very capital-intensive and the foreign oil monopolists can, if they choose, keep the world price just below the level that would make an investment in shale and liquid fuels from coal financially viable. In fact, one of my colleagues at M.I.T. suggests that this is, in fact, OPEC's pricing policy. If this is true, perhaps it would justify the cost of synthetic fuel programs just to hold down their prices; at any event, it is an interesting point.

I hope that government and industry together will find a way to buffer synthetic and shale projects from these uncertainties, because they have a great deal of future. In fact, I believe that if America's economic survival depended upon it (as it may), the country could be producing as much as three and maybe even five million barrels of oil from shale and other synthetic fuels within a decade or a little longer, and possibly competitive with today's oil prices, and certainly cheaper than prices that will exist within the decade if the recent rate of increase for petroleum continues.

Aggressive Conservation

To meet such an ambitious goal the nation really must have the will and determination. We would have to eliminate many of the existing road blocks, including some of our valued environmental standards. We could, I think, rather quickly use a great deal more coal directly for electricity and industrial production, and it would not cost very much more. What is required here is merely to remove the existing uncertainties regarding pollution control requirements; please note that I said, remove the *uncertainties,* not the requirements. "Get the noise out of my learning system," in other words. Over the same decade, an aggressive conservation program could save the equivalent of at least two to three million barrels of oil, split between industry and consumers. Industry has actually done very well in conservation and continues to do so, but the public has not yet responded to the degree that is possible, and I think there is a real problem of convincing people of the need to conserve.

In all of this, and particularly in the area of conservation, it is possible to see the difficulty of getting the replacement technologies I mentioned earlier into play. Approximately a hundred million cars on the roads today cannot be improved; they can conserve fuel only by being driven less. It will take many years to replace the inefficient ones presently on the highways.

To do the various things I suggest are possible—to meet size criteria, develop adequate and not too expensive fuels—to solve the short-range problem (which is, I think, the critical one for the United States) would require major investment, perhaps twenty to thirty billion dollars a year. Most of it I hope would come from the private sector. This is not a large amount of money compared to the federal budget, or even to the U.S. payment deficit. Of course this does not include the cost of new automobiles, for which Americans now spend more than $50 billion a year, just to give a standard of reference. I would also recommend taking whatever initiatives public opinion makes possible to increase the availability of nuclear power, which in my judgment remains an important ingredient of our energy future.

Optimism for the Long Term

Looking beyond ten or twenty years into the next century, we in the United States can feel quite confident about the availability of raw materials. As we know, the coal deposits here can be a major energy source for hundreds of years, and a continuing source of liquid fuels. Fuels from shale will continue to be available for at least a half-century. Nuclear energy, in spite of its current problem, will and should remain a significant part of our energy system. Solar energy will become more effective, and fusion energy should become available along the way; in fact, my technical intuition tells me that in the long run the world will obtain much of its energy from these two sources. And over all, energy will doubtless continue to become more costly each year and have to be used more judiciously.

All of this is to say that we should be quite hopeful. The United States has the basic energy resources to keep our society going. Furthermore, the technologies needed to make those basic resources usable are available. We have enough industry that has done equally difficult things before, capable of learning how to develop and apply the energy we have.

But America is also saddled with a society too constrained to be able to learn adequately, or to react without serious dislocations, so the fundamental energy problem is really a communications problem. The vital task is to get enough of our fellow-citizens and particularly our policy-makers to understand the options we have, and then to arrange matters so that the energy industry can get on with the production of these harder to make, but essential, fuels.

Discussion

Dean Rusk, former Secretary of State (from the floor): I must confess that I am one of those who simply do not believe that fossil fuels as we know them are present in this thin crust around the earth in quantities sufficient to satisfy the rampaging appetites of a growing population indefinitely into the future. I am concerned because I suspect that before my children, and certainly my grandchildren, get to be my age, one of the oldest causes of war will be revived in a nuclear world where thousands of megatons are lying around in the hands of frail human beings. I am referring to the pressure of people on resources.

We all saw during the 1960s an extraordinary combination of management, science, technology, and production called NASA, which was assigned the problem of putting somebody on the moon. It was an effort that called upon hundreds of companies, hundreds of universities. It was a brilliant job from the point of view of human organization, purpose, and so forth. And it caught the imagination of the public. What would you think of an attempt by the OECD countries—North America, Western Europe, and Japan—to put together resources and brain power and engineering and production for a comprehensive attack on the energy problem, not less in scale and scope and urgency than the space program?

You talked about new technology. Part of my question, of course, is, are you satisfied with the pace at which that new technology is being pursued? Do you see any possible benefit from a strong international effort so that we Americans, among other things, begin to recognize that we are all part of a family of man and we go down together on this, or survive together?

Wiesner: I could write a book to answer that question. I have been limiting my remarks to the theme of this conference, namely the problems of the United States. I certainly share Dean Rusk's concerns about the longer-range future, particularly about the population explosion. Every time the population doubles, the resources available world-wide halve—unless something has been done to keep up. I made a calculation the other day that I hope was wrong; I figured that each of us has available to us on the surface of the earth at the present time a total of seven acres. That is not much earth resource. And if population doubles and doubles again, we shall each have two acres, and that is a rather sparse amount of land on which to survive. I think the number one problem in the world is the population explosion; and closely second is the nuclear weapons proliferation. People worry about the dangers of a nuclear power plant like Three Mile Island, and they should, but any day *millions* of people could be killed by an accidental explosion of one nuclear weapon, if it happened to be the right nuclear weapon. And yet the great powers go on in a mad arms race, wasting resources, and living so dangerously.

I personally doubt (and this is a gut feeling, not a scientific fact) that unless population growth can be controlled, technology will ever be able to manage. Another consideration here is that historically it is known that population growths slow down when people become more prosperous, and particularly when women become more educated and have outlets other than raising families; and so it is an important part of our own future to try to help the development of other countries.

With regard to technology, I would say not enough is being done. Industry is doing reasonably well, but it has relatively limited resources. Government, however, is not organized to deal with technical things. It tends to manage development programs the way it manages basic research programs, which is to give central direction from Washington and farm out little pieces here and there. A program of any scope ought to have everyone working on it together in one location and with management only peripherally from a central point. This has been destructive of development in almost every part of government, the Defense Department and NASA excepted.

I think it is a mistake to use the analogy of the space program in talking about these much more complicated societal problems. The space program had a single purpose in mind—to get from here to there and back once or twice, a few times—and the system was never taken out of the hands of the people who designed it, the scientists and engineers. In fact, when Apollo 13 got into trouble, the scientists in our laboratory at M.I.T. took over by communication, and the designers of the system were able to figure out and tell the people in the spacecraft what had gone wrong. At the time of the accident at Three Mile Island, the designers could not be found; they had probably long since left the employ of Babcock Wilcox; at any rate, they were not available to answer questions and solve problems.

In the short-range energy area, I believe more technology and much more know-how, not only in the technical sense but in the commercial and industrial sense, is available now than is being exploited. I was focusing on why the technicians could not get on with that set of problems, but what I have said applies equally well to longer-range problems, if not more so. Obviously it is harder to bring one's attention to the long-range picture and worry about the year 2000 if the problems of today cannot be solved.

From the floor: Was it perhaps learned from the Three Mile Island accident that the best future alternative energy source is solar, not nuclear? That has been suggested in principle by the reaction within the public as a result of feedback from the accident. The implication is, we do learn as a society, and soon enough to survive in the long run.

Wiesner: I would quarrel with most of that statement. It would be fortunate if there were a solar energy technology that could be exploited other than marginally for some heating—home heating and water and space heating in a few areas—but the fact of the matter is that in spite of enormous effort, work, research, it is not known how to use solar energy competitively. For example, to make electric power by solar energy today would cost twenty times as much as using either fossil fuels (coal or oil) or nuclear power; and it is a great pity, for there are many advantages to solar energy. There are also some disadvantages to it. One real problem is that solar energy is very diffuse. At the peak of the solar cycle in a place with lots of sun and no clouds, one gets about a kilowatt of energy per square meter. To capture even 20 percent of that, one needs five square meters of collective energy per kilowatt or 5,000 square meters per megawatt. That suggests how large a "farm" of these things would be needed, even supposing one knew how to make them for a thousand megawatt equivalent.

Nonetheless, many research institutes, including the one with which I am associated, are working very hard to find the invention, the discovery, the technological breakthrough that will make solar energy practicable. No-one knows when it will happen—perhaps soon, perhaps not for a long time. So it is a mistake for people to say the choice is solar instead of fossil fuel, or solar instead of nuclear. Our society does not have a potential solar energy economy except at a cost that would, in my view, be very destructive to our standard of living.

From the floor: Why not have the government contract long-term to buy two or three million barrels of synthetic fuel at a break-even price?

Wiesner: If the government were willing, that would be a great thing. I would be all for it. To try in a decade to get up to three million barrels a day of a combination of synthetic fuel from coal and shale would require twenty or thirty independent sites at which the systems would be permitted to grow. Maybe one should establish a kind of net price, a system price, that the government would be willing to pay; people who beat that price would make money, and people who did less well would lose money, so there would be an economic incentive. There are many ways of doing this.

From the floor: In discussing the need for tax-plus regulatory innovation, you suggested something like the government providing risk capital as in West Germany, and standing back to let free enterprise do the job. Do you have ways to allocate the benefits out of a successful undertaking such as this equitably?

Wiesner: I have thought about it a lot; I cannot say I have a way. I have actually tried to make a mathematical model of this process based on discounted return on investment. Sophisticated replacement technology such as I have been discussing takes ten, fifteen, even twenty years to get one's money back; ask whether this is a good private investment and the answer is always no. A private investor is much better off, particularly in the face of all the regulatory uncertainty, to put money in short-range things, where it is possible to make some prediction and where the pay-off comes soon.

From the floor: If we are becoming a nation of complainers as you suggested, aren't university students and recent graduates valued? Why is this, and what can universities do about it?

Weisner: I do not know, but I have observed that students and young people in general reflect older people in the society. Yankelovich, Skelly and White did an interesting poll last year of both young and old people, and found roughly the same attitudes in cross sections—a desire for security, a growing dependence on the government, a sense of alienation from the institutions to which they belong, including companies they work for—that were fairly pervasive in both groups. I think this is a very serious problem in our country, something that has to be reversed as part of creating a more hopeful society.

Arthur M. Bueche

Energy— An International Appraisal

The question is, "Can America prosper in an era of plentiful although higher priced energy?"

Let me make my own beliefs clear on a couple of points. I am not all that confident about the *near* future; but yes, it *is* possible, though certainly not a sure thing, that America can have the energy it needs for the long-range future. And most certainly, yes, that energy will be higher priced.

But will America prosper? Will we as a nation succeed? become stronger? flourish? That is another question. The fact is, whether we prosper will be a relative matter. Prosper compared to what? Or to whom? How well we actually do economically in the years ahead will depend to a significant extent on how well our international competitors fare. How will energy supplies and energy prices affect them?

Similarly, both we and our competitors in world trade will be vastly affected by how energy supplies and energy prices impact on our mutual customers, many of whom are in the developing nations. Will they be able to buy?

Arthur M. Bueche *is Senior Vice President for Corporate Technology of the General Electric Company. He is a member of the Executive Committee of the Council of the National Academy of Engineering and of the Academy Forum Advisory Committee of the National Academy of Sciences. Dr. Bueche is a past president of the Industrial Research Institute and is or has been an adviser on technology to many government agencies, universities, professional societies, and business organizations.*

Incidentally, we Americans are, after all, only human. We share fully in mankind's frailities, lusts, and jealousies. Thus, how well we prosper will, in many ways, depend on how well we think we are prospering. "Will we have the good life?" is a good question. Perhaps at least as important is, "Will the good life be good enough to live up to our expectations?"

A Statistical Matrix

As a start about what I might say at this conference, I created a great and grand matrix that might serve as the basis for my paper. This matrix, I told myself, would truly put into perspective the whole matter of how "prospering" depends to a significant extent on comparisons with how well others are doing.

Across the top of my matrix, I listed the major nations, areas, blocs of the world. Down the side: the key elements of "statistical energy"—such things as domestic production of energy, exports, imports, ratios to GNP, energy use per capita, costs trends. It all seemed properly designed and neatly aligned—that is, until it came time to fill in the three hundred vacant spaces.

Some of the results of the search for figures that ensued are included in a simplified version of my originally conceived super matrix, which is appended to this paper.

I should sound a certain note of reservation about the numbers here assembled, and perhaps it can serve also as a caution to all those who would seek to solve the world's energy problems by stockpiling statistics.

1. Global data on energy, on anything even remotely resembling a timely basis in a fast-changing world, is often difficult if not impossible to obtain.

2. No two sources of such data seem to use the same units of measurement. The available graphs, charts, and tables are a maze of Mtoe's, barrels-per-day, quads, terawatts, cubic meters, kilocalories, and what-have-you's. In the appended matrix, all possible units have been converted—in the interest of uniformity—to quads; that is, quadrillion Btu's. The quad is a very large unit. One quad provides the present needs of three million Americans for an entire year. (Incidentally, establishing some means of getting better and more consistent world-wide energy statistics might be a worthy recommendation of this conference.)

3. Different sources often provide quite different numbers, even though both may flaunt their accuracy with an abundance of significant figures, frequently extending far right of the decimal.

4. It is possible to generate even more—and more imposing—numbers to the right of the decimal, thanks to the uncanny ability of today's pocket calculators to divide one approximation by another approximation with astounding accuracy.

5. One can easily get so engrossed by, and lost in, the surfeit of numbers that he completely loses sight of the question at hand.

That question is, as I recall, "Can the United States prosper in an era of plentiful although higher priced energy?" My hope is that I can make a contribution to this discussion by outlining the present U.S. energy situation vis-à-vis that in other nations, and do this in the simplest possible terms.

The objective of this exercise, of course, is to see if any educated guesses can be made about how these international relationships may or may not shift with changes of energy supply and price during the years just ahead. Then we can hypothesize about how these results will affect America's prosperity (or, at least, how we view America's prosperity). And, of course, we can hope that we may develop ideas for the kind of action that can help nudge things in the direction we want them to go.

International Competitors

As I see it, from an economic (that is, world trade) viewpoint, the U.S. has two major competitors: Western Europe (the EEC) and Japan. Politically, we are mainly concerned with the relative global strengths of Russia and China. Then I suggest we look at Canada and Mexico, not in terms of their being competitors but as immediate neighbors and friends who, inevitably, must share our nation's strengths and be affected by our problems.

I realize that this leaves out huge areas like India, Indonesia, South America, and Africa, to say nothing of the OPEC nations themselves. But there *is* a limit to how many statistics any one of us can swallow, and besides, I think I can make my point without them.

Further, for the sake of simplicity, I will try to avoid all those confusing units of measurement. We are not really going to measure; we are just going to compare. So, in most cases, on the graphs included in this paper, the U.S. situation (either now or during the most recent year for which corresponding figures for the other countries are available) are designated as one—1.0.

How this system works is readily apparent on this simple graph showing the *relative* **Population** of the seven nations or blocs that we have selected.

Population (US = 1.0) 1976

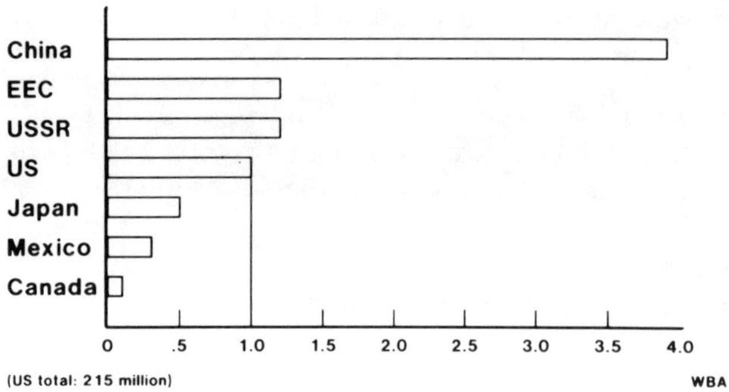

I could tell you that this means that, in 1976, China had an estimated 836 million people, the European Economic Community had 259 million, Russia 257, and so on. Instead, at the lower left corner of each graph a single number will be found, for the U.S., which may help establish a mental base for the quantities we are comparing.

(The relative lengths of the bars, as analogs, are probably at least as accurate—although not as presumptuously accurate—as would be more detailed digital guesses at the current figures.)

For each chart, the elements are in ranking order, which means that the United States is often (but not always) number one on the top—as is the case with the chart showing relative gross national product **(GNP).** (The sources of the data on which these charts are based include the 1978 World Bank, the United Nations, the International Energy Agency, and many others. They are identified on the matrix in the Appendix. Some of the figures are based on unclassified, published reports of the CIA.) It should be noted, in viewing this chart, that recent gyrations in monetary exchange rates have served to complicate the comparisons of GNP data.

The majority of the statistics I have included will be shown in "per capita" terms, as in the chart headed **GNP/Capita.** It seems to me that economic prosperity—which is our subject—is something enjoyed by individual people. Admitted, such comparisons, like all others, may be

GNP (US = 1.0) 1976

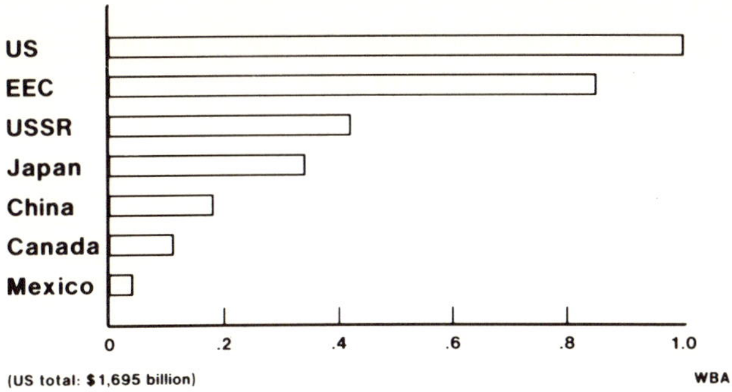

GNP/Capita (US = 1.0) 1976

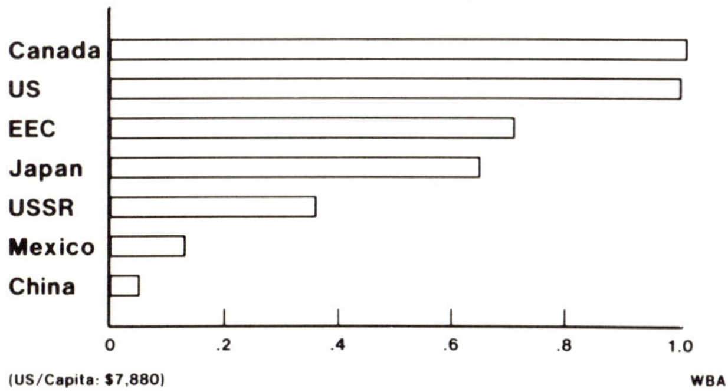

vulnerable to a number of caveats, but this approach seems to offer some insights into how the various peoples of the world are faring in a way that often is not apparent when the usual national or area statistics are used.

As a lead-in to energy comparisons, then, let us consider some other per capita relationships. In the chart, **Exports/Capita,** Canada is seen to be at the top; it is, of course, a large and rich area with a relatively small population. (Mexico is included on most of our charts, but its numbers were not available for this particular comparison.) The EEC bar is misleading as shown. It includes internal exports, analogous to our interstate commerce. For a fair comparison, it should be cut almost in half, making it about the same as the U.S. and Japan.

This is also true on the chart showing **Imports/Capita.** Both Exports/Capita and Imports/Capita have been included only as a basis for the chart showing **Net Trade Balance/Capita,** which emphasizes an ex-

Exports/Capita (US = 1.0) 1977

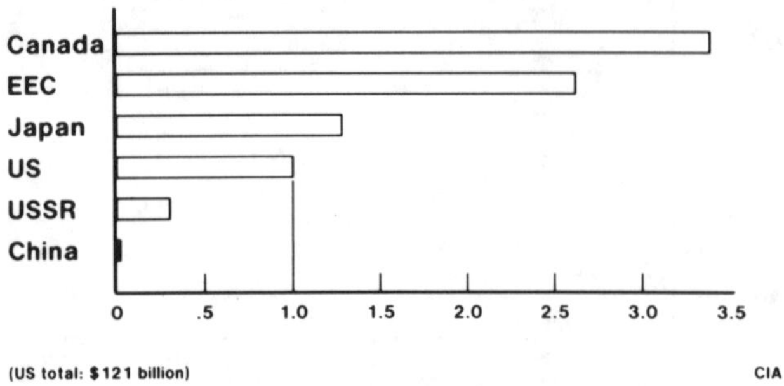

Imports/Capita (US = 1.0) 1977

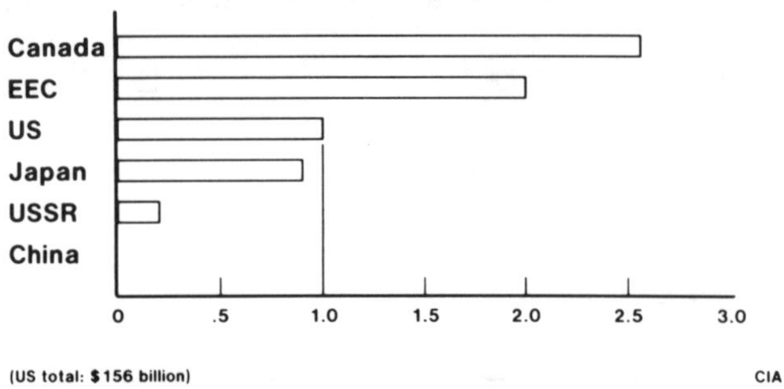

Net Trade Balance/Capita ($'s - US) 1977

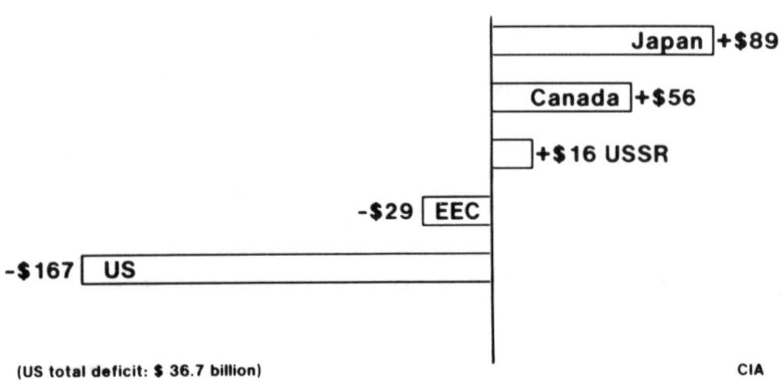

tremely important point that is all too familiar. The U.S. balance-of-payments situation is dramatically displayed. The big plus-nations are, of course, those in OPEC, not included here. Obviously, the comparisons made on the Net Trade Balance/Capita graph deserve rather special attention. Remember that the negative figure of $167 for each one of us is for just that one year.

Crude Steel Output/Capita (US = 1.0) 1977

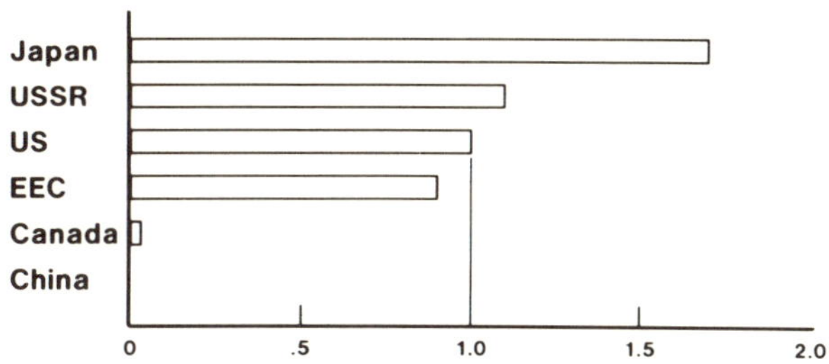

(US total: 113 million metric tons) CIA

Cement Output/Capita (US = 1.0) 1977

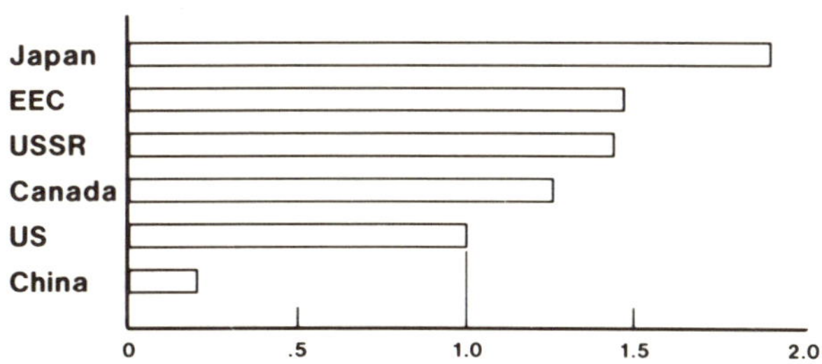

(US total: 73 million metric tons) CIA

Grain Production/Capita (US = 1.0) 1977

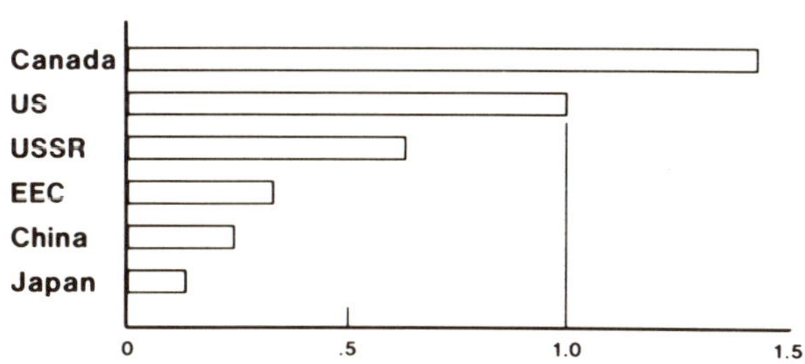

(US/Capita: 1210 Kg) CIA

Meat Production/Capita (US = 1.0) 1977

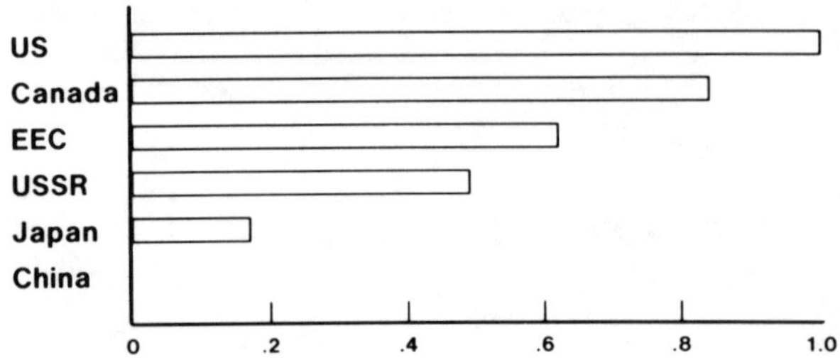

(US/Capita: 117 Kg) CIA

It is worthwhile to examine some key items of production in a per capita relationship: **Crude Steel Output, Cement Output, Grain Production,** and **Meat Production.**

Japan's preeminence in the production of steel is well-known. Note also, however, the strong position of the Soviet Union. I can account for the U.S. position with regard to cement output only by suggesting that we have already built most of our highways and are now down mainly to filling in the pot-holes.

People who gather numbers to compare countries, societies, and cultures have some favorite ways of seeking to measure affluence, if not influence. Three of these are **Radios/Capita, TV Receivers/Capita,** and **Telephones/Capita.**

It is not surprising that the rest of the world trails the United States in these categories. There now are two radio sets for every American man, woman, and child in the United States, and similar numbers for TV receivers.

The proximity of the per-capita figures on TV receivers for Japan and

Radios/Capita (US = 1.0) 1974-76

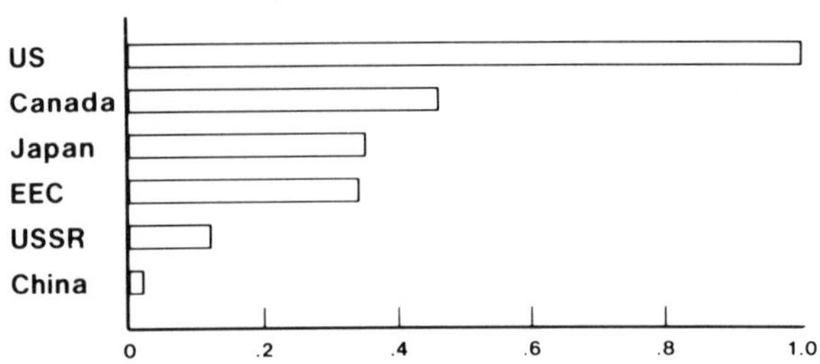

· West Germany only
(US/Capita: 1.98) CIA

TV Receivers/Capita (US = 1.0) 1974-76

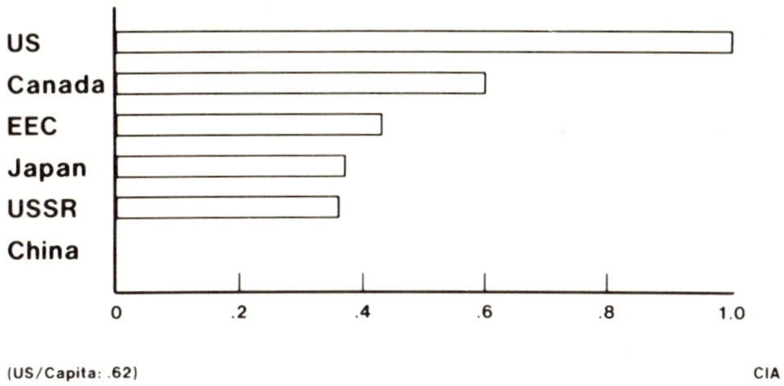

Telephones/Capita (US = 1.0) 1977

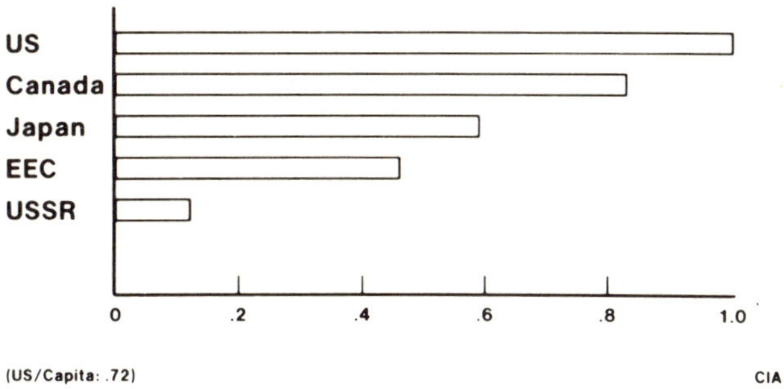

Russia may appear questionable, but I am reluctant to suggest to our CIA friends who provided these numbers that they go back and re-check the Russian numbers, apartment-by-apartment.

As a final "background" comparison before going on to energy statistics, the **Consumer Price Index** graph measures relative inflation,

Consumer Price Index $\frac{1977}{1970}$

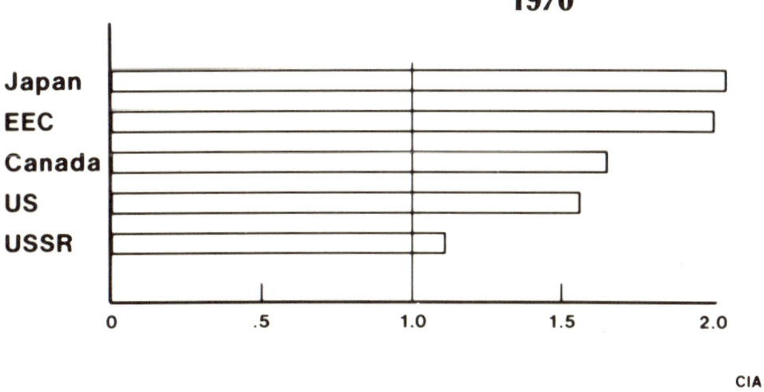

1970 through 1977. As can be seen, in spite of all this country's problems and concerns, it had not then suffered as much as our two major economic competitors. Figures for Mexico and China are not available, and I am told that the numbers available from the USSR have relatively little meaning.

Now let us move to per-capita comparisons of energy statistics, and

Domestic Energy Production/Capita 1976
(US = 1.0)

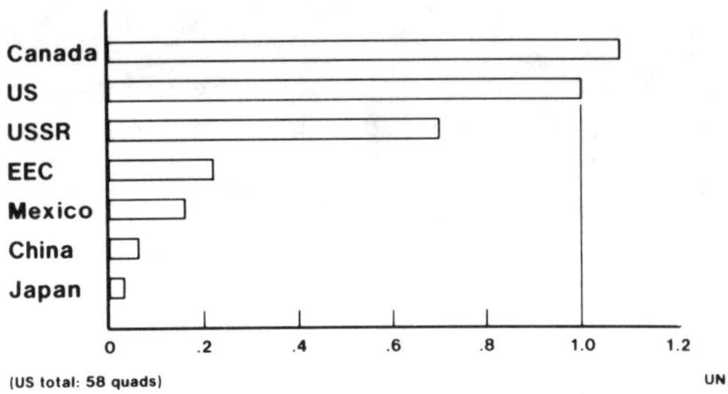

start with **Domestic Energy Production/Capita,** the energy made by each nation using its own fuels. Japan's utter dependence on the outside world is particularly apparent.

Energy Imports/Capita (US = 1.0) 1976

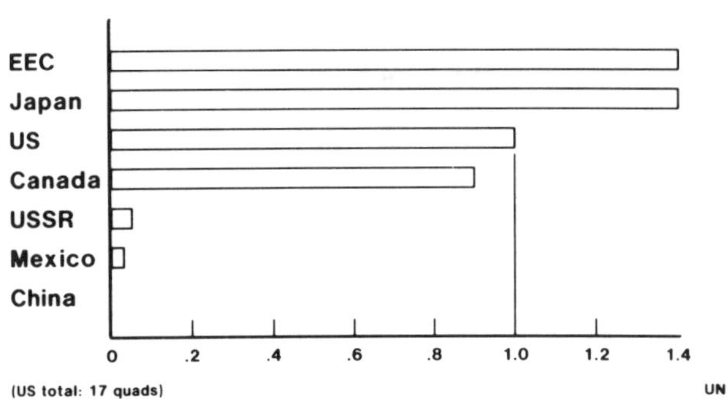

This is reflected directly, of course, in **Energy Imports.** Although Russia is shown to import some small amounts of coal, oil, and gas, it actually exports some eight times as much energy as it imports.

Energy Exports/Capita (US = 1.0) 1976

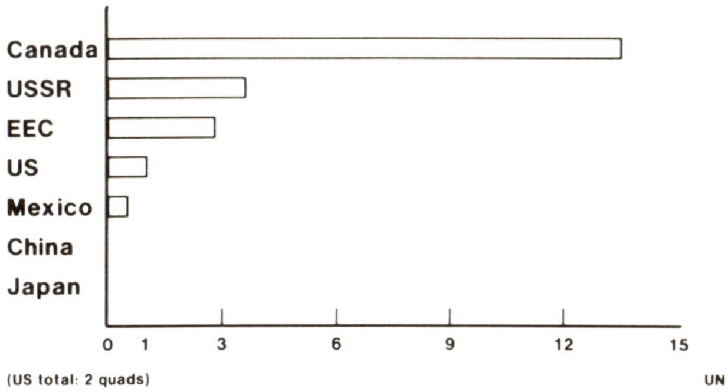

In regard to the **Energy Exports** chart, it should be emphasized that none of these countries actually exports very much energy. Even the top bar in the Exports chart would shrink to virtual invisibility if we were to change the bottom scale to permit inclusion of the OPEC nations.

Net Energy Balance/Capita 1976
(Barrels of Oil-equivalent)

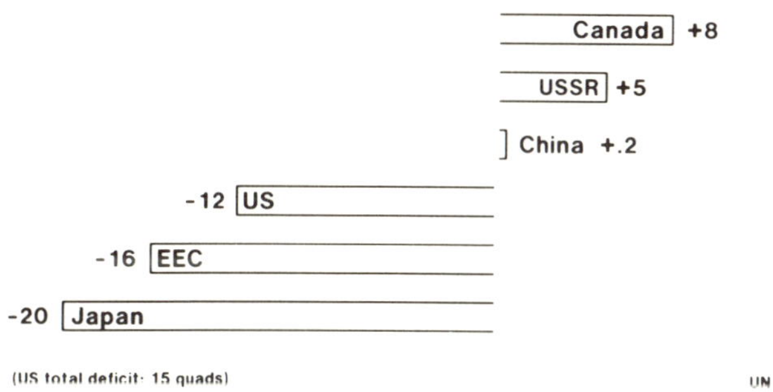

The **Net Energy Balance/Capita** chart represents a sort of "balance of payments" for energy. The United States deficit is only 12 barrels of oil, but that is 12 barrels for every man, woman, and child. The even greater dependence on OPEC of the people of Europe and Japan is graphically indicated.

Combining domestic production with the plus or minus of exports/imports, one arrives at the numbers for **Energy Use/Capita.**

In many ways, these are the most controversial energy comparisons on the world scene today. They raise the obvious question: "Why do the people of the United States require so much more energy per person

Energy Use/Capita (US = 1.0) 1976

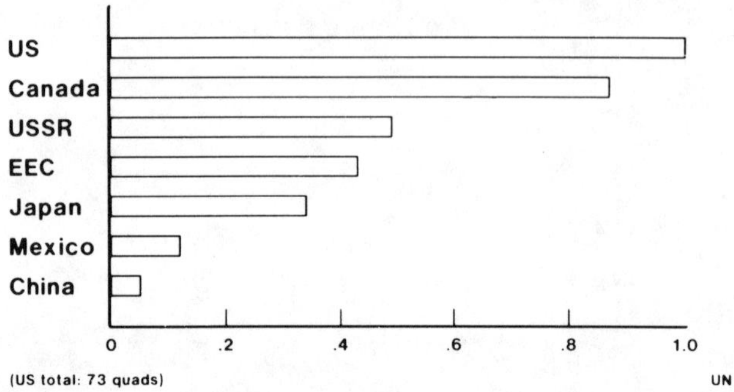

than even relatively prosperous countries in other parts of the world?"

One part of the answer, of course, is our geography and geographical distribution to our homes relative to our places of work. We devote an inordinate amount of our energy, by comparison, to transportation. Also, we devote substantial energy to the task of being principal defender of the free world. Perhaps the most important way to view the chart, Energy Use/Capita, is to suggest it means the United States should be able to compete in an era of higher-priced energy because it has so many more opportunities than its competitors for conservation and belt-tightening.

Energy Use/GNP (US = 1.0) 1976

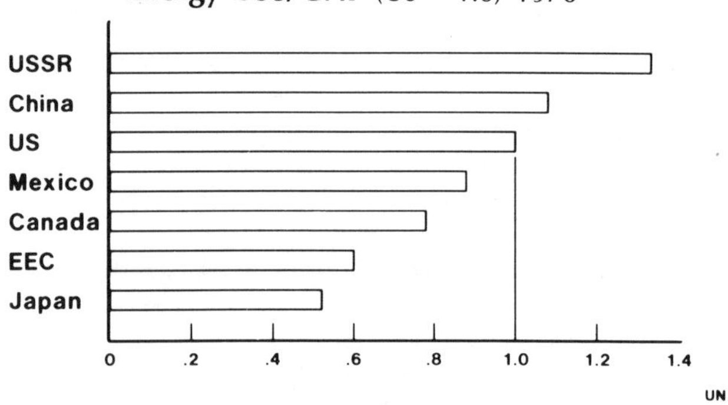

The chart entitled **Energy Use/GNP** is probably the most tricky of all in this paper. It must be viewed with great care. The history of energy-vs.-GNP relationships in this country has been extremely complex, especially as we have changed the mix of fuels commonly used. However, it should be noted that the ratio, with a few exceptions, has declined more or less steadily over the past hundred years. Even so, the U.S. number still amounts to an amazing 7.6 barrels of oil equivalent for every thousand dollars of GNP.

We turn now to **Oil Production/Capita.** Canada really stands out here, and I expect Mexico will, too, in a few years. Note that the USSR and U.S. are about even.

Oil Production/Capita (US = 1.0) 1976

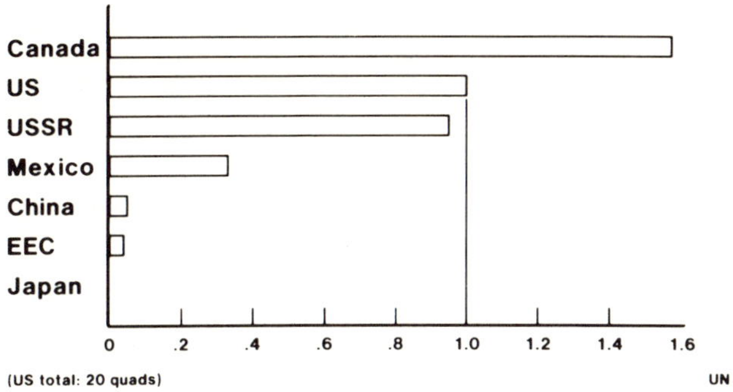

But look at **Oil Imports/Capita.** The Russian figure is so small it does not show. As might be expected, Japan and the EEC are at the top of the list, and their per capita figures are remarkably similar.

Oil Imports/Capita (US = 1.0) 1976

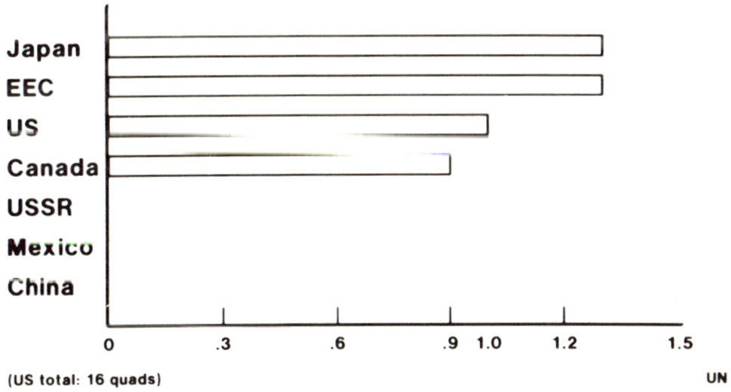

Next, **Oil Exports/Capita.** The Soviets are seen to export more than 6.0 quads of their 22 quad production. Their net use of oil is about one-half of ours. Here, again, the EEC figure is misleading because it includes internal exchanges.

Oil Exports/Capita (US = 1.0) 1976

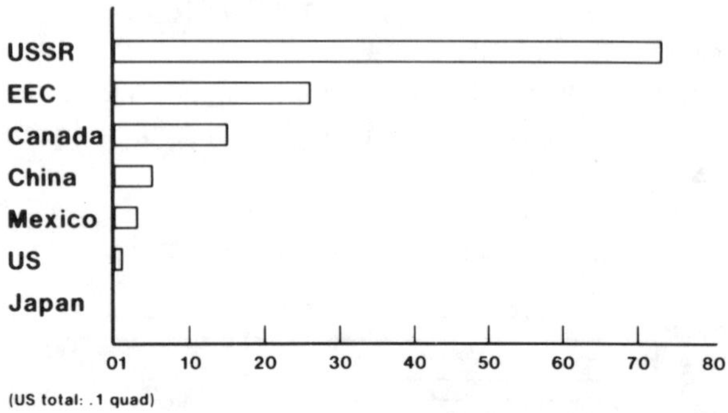

Perhaps for a discussion of future trends, the numbers on **Estimated Proved Oil and Gas Reserves/Capita** are more pertinent. Once again, remember that these are per capita comparisons, and as such they may not give full weight to the importance of being able to provide energy to customers and allies. They are, nonetheless, revealing. These numbers include most—although perhaps not all—of the recently confirmed resources in Mexico. Some recent claims would move the bar on the chart out perhaps 50 percent or more. China is low today, but I predict a great deal more will be "proved" there before long.

Estimated Proved Oil and Gas 1976
Reserves/Capita (US = 1.0)

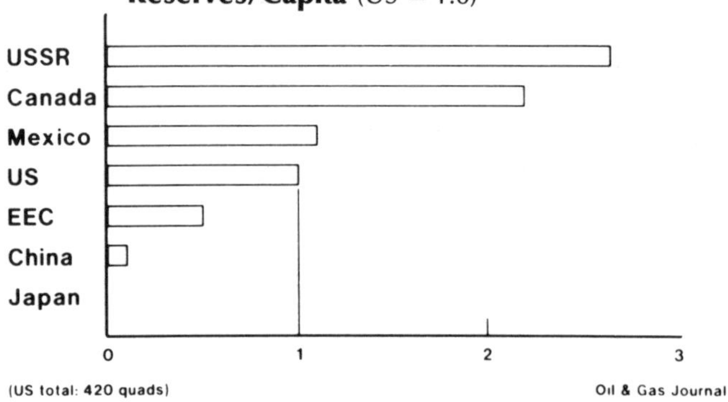

The great U.S. strength, as far as fossil fuels is concerned, is, of course, coal. Most people think the figure of 6,000 quads of reserves for the United States is quite conservative. On the chart, **Recoverable Coal Reserves/Capita,** note that Europe also still has significant coal reserves, as does the USSR. I need not remind this audience of all the economic,

Recoverable Coal Reserves/Capita (US = 1.0) 1976

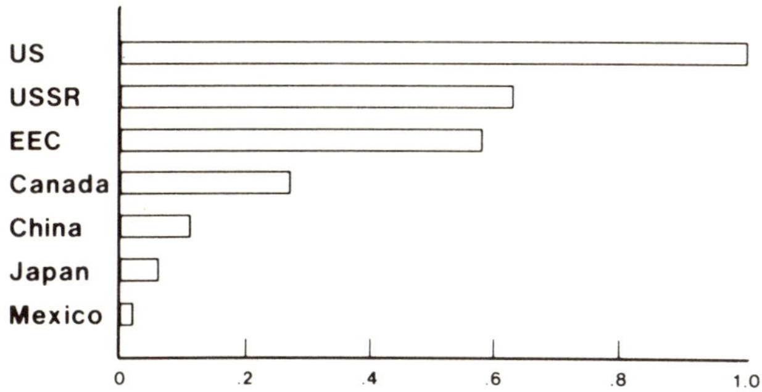

(US total: 5840 quads) UN

environmental, sociological, and technical questions that revolve around the possible future use of coal resources—especially in the United States.

Thus, when combining the previous two charts, as is done in **Estimated Proved Fossil Fuel Reserves/Capita,** it is important to remember that well over 90 percent of the United States' fossil reserves are coal. The numbers seemed too hypothetical to put on a chart, but if you assume that all these reserves can be used, and if the countries would be wholly dependent on them, *and* if present levels of consumption would remain constant, even then there is only enough energy here to keep the U.S. going for about 85 years, the USSR for perhaps 140, the European Community for maybe 110—and Japan for a dozen years at most.

Estimated Proved Fossil Fuel Reserves/Capita 1976
(US = 1.0)

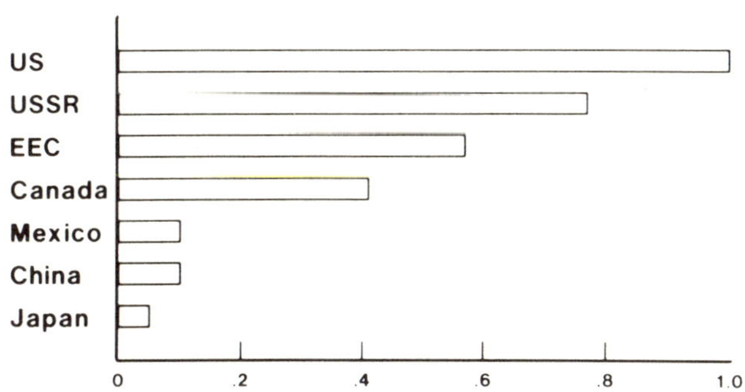

(US total: 6260 quads) Oil & Gas Journal; UN

Uranium Reserves/Capita (US = 1.0) 1976

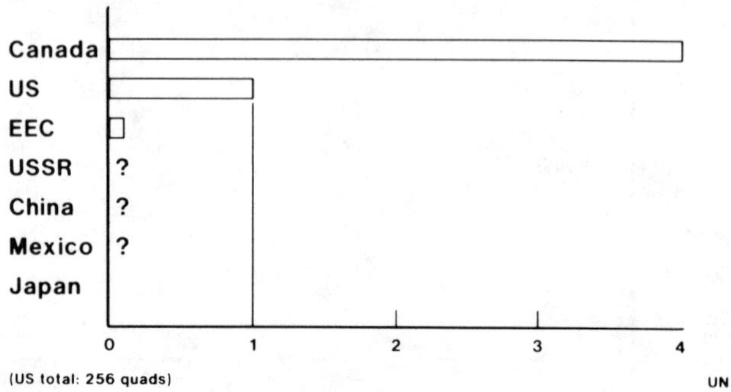

The figures on **Uranium Reserves/Capita** are incomplete. At $30 a pound, the United States seems to have a tangible portion of the world's reasonably assured uranium reserves; they are estimated to be worth 256 quads, using present-technology light-water reactors. By the same reckoning, Canada has 90 quads (it eclipses the United States on a per capita basis, however) and the EEC perhaps 40. Advanced breeder technology would, of course, make a big difference in the value of these resources. Although not shown here, most of the world's better uranium ore is in Australia, South Africa, and Sweden, each of which has approximately 150 quads.

Canada's vast use of hydro is reflected on the chart showing **Electricity Production/Capita.** Canada's hydro output nearly matches that of

Electricity Production/Capita (US = 1.0) 1976

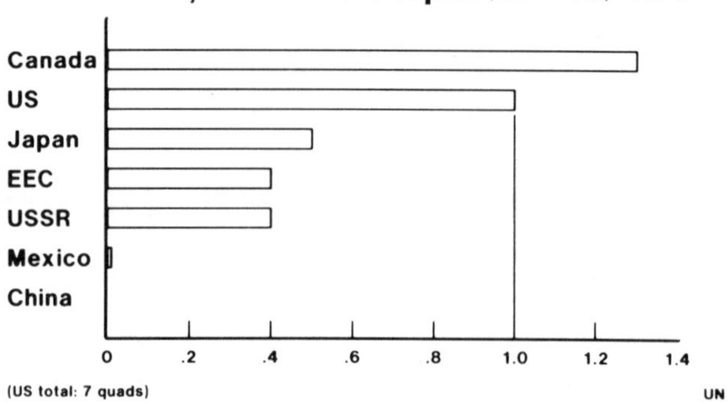

the United States—for only one-tenth the number of people. Noted at the lower left is the fact that U.S. electricity production is approximately 7 quads. Some 23 quads of fuel energy were consumed to generate this electricity, which helps underscore the importance of the many continuing efforts by electrical equipment manufacturers to keep steadily improving the thermal efficiency of central-station generating plants.

Canada's abundance of hydro-power sites also accounts for the low

Price of Electricity (Home) (US = 1.0) 1975

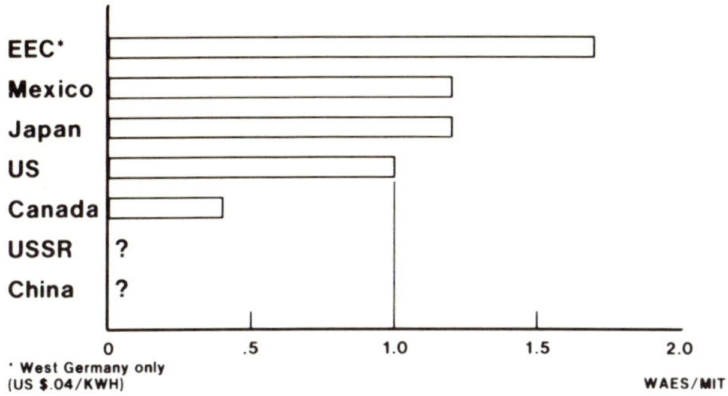

Price of Electricity (Industry) (US = 1.0) 1975

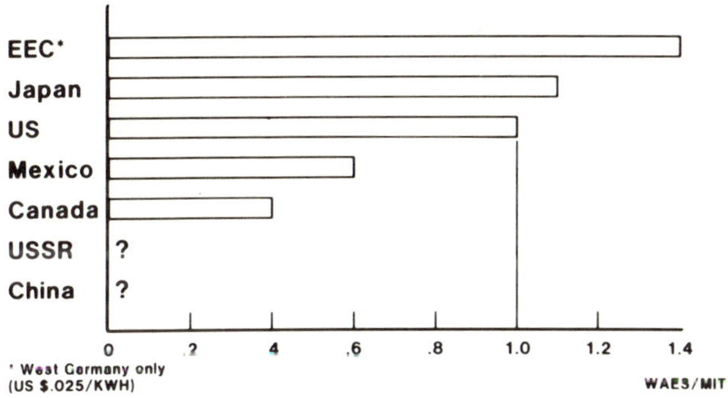

cost of its domestic electricity, shown on the chart, **Price of Electricity, Home.** Europeans (although in this particular case the figure shown is only for West Germany) pay somewhat more for the electricity they use at home than we do, but the differential in the cost of electricity provided to industry **(Price of Electricity, Industry)** is not as great.

Of greater current interest is the relative **Price of Gasoline** for automobiles. All of us have more up-to-date information from first-hand experience than these mid-to-late 1975 figures, but I would surmise that the ratios have not changed all that much. With all our concern about higher gasoline prices, we will likely get little sympathy from our Ja-

Price of Gasoline (US = 1.0) 1975

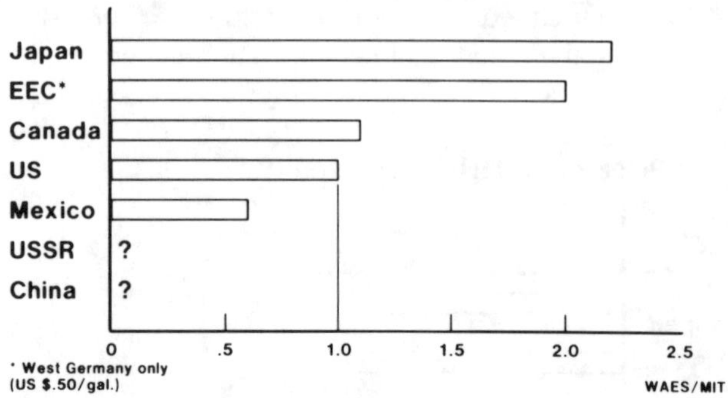

panese and European friends who have been paying a much higher price than we for quite some time.

The final three graphs in this paper relate to how the United States—and its competitors and friends—now *use* the energy available to them. Unfortunately, the only available figures for other nations are for non-electric energy consumption, but the comparisons are still well worth

Non-Electric Energy Consumption - 1976
Industry/Capita (US = 1.0)

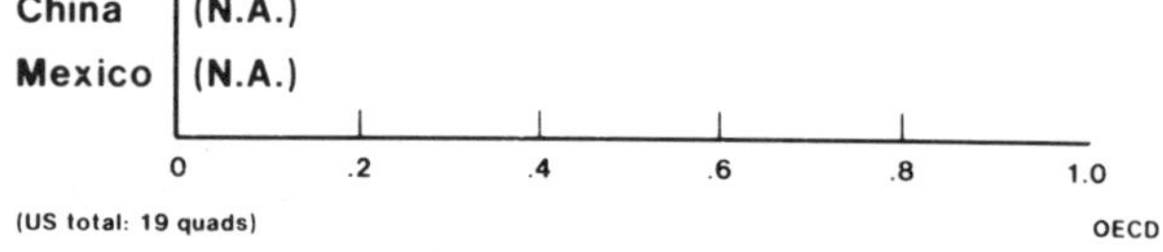

making. First is **Non-Electric Energy Consumption in Industry/Capita.**

Japan's *total* use of energy per capita was shown to be about one-third of ours; for industrial purposes, Japan's use is closer to a half of ours. What would be a particularly significant comparison is one for which it seems especially hard to find good numbers: the relative industrial output per industrial energy input for each country.

No across-the-board breakdown between residential, commercial, agricultural, and other energy uses is available, but the chart, **Non-**

Electric Energy Consumption—Residential, Commercial, etc./Capita, would seem to reflect differing needs and habits in regard to space heating in particular climates, and to living standards. I remind you that

Non-Electric Energy Consumption - Residential, Commercial, etc./Capita 1976
(US = 1.0)

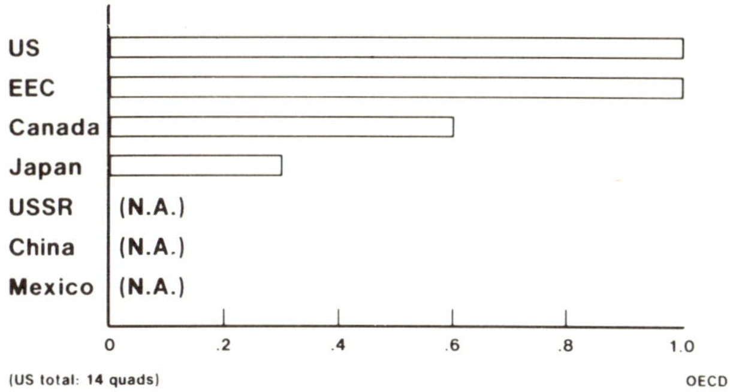

these figures are for *non-electric* energy consumption, so when we speak of the needs of the climate and living standards, we should note that airconditioning in the United States is not generally included in this comparison.

The most striking difference is shown in the final chart, **Non-Electric Energy Consumption—Transportation/Capita.** Although the United States surely has broad opportunities for conserving the use of energy in industry and at home, our major target will have to continue to be transportation. The numbers speak for themselves: four-to-one over

Non-Electric Energy Consumption - Transportation/Capita (US = 1.0) 1976

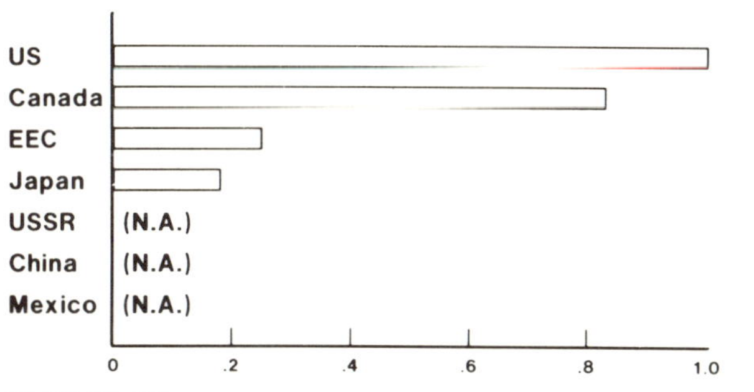

Europe, five-to-one over Japan, and probably ten-to-one over the USSR.

Those, then, are some per-capita comparisons about energy and its use, made in the most straightforward kind of terms the available statistics permit. At this point, let us remind ourselves that our hypothesis presupposes that in this country we will have energy available to use—if we can and will pay for it. The question is: *Can we do so and still prosper?*

In this regard I am struck first of all by the fact that there are very few energy-related factors where the United States has a position inferior to that of its strongest present trade competitors. Compared to Japan and the members of the EEC, I see no reason why the U.S. should not get along quite well—by comparison. Our present energy prices are rising, but they are still lower than what the people of these other countries have to pay. And although energy prices have higher public visibility in times of inflation than many other costs, their actual impact on the Consumer Price Index is lower than most people would guess. The cost of food has three times more impact on the CPI than does the cost of energy. Also, the United States surely has more opportunities for energy conservation than have most other societies.

In other words, if the United States does not seem to prosper in the years ahead vis-à-vis its chief free-world trade competitors, it will be because of reasons other than the higher cost of energy.

The USSR and China

What, then, about the big-picture arena of global political power?

In what might be called the "match up" of the free world versus the USSR and China, the most serious concerns for the question we have been considering are resolved by our prescribed hypothesis: we *will* have sufficient energy in the United States. For now, at least, this supposition eliminates a lot of unpleasant possibilities about which we otherwise might have to worry.

Viewed this way, these giant blocs must be considered on virtually the same basis as our trade competitors—as potentially very large factors in international markets.

Here, however, some of the comparisons are somewhat less reassuring. Both China and the USSR are, at this time, in the plus column for Balance of Payments—not by much, but still plus. Both are also, for the present, energy self-sufficient. Russia has vast proved resources—far more oil than we, and nearly as much coal. China's proved reserves may be low, but who is to doubt the hidden potential of that huge land mass?

Meeting Expectations

Early in this paper, I mentioned that Americans will prosper to the extent that their expectations are met. The people of the rest of the world react the same way. In the controlled economies, living standards can be allowed to increase slowly, although fairly steadily, maintaining at least a semblance of public satisfaction. Many view this approach as the basis for Japan's remarkable growth as a major force in world markets. The Russian and Chinese leaders can be expected to play this game to the hilt. By thus controlling domestic consumption, they can allocate more energy to industry and produce at far lower costs than nations whose peoples' aspirations and expectations look upward from a far higher base.

In light of the kinds of comparisons that have just been made, it would seem that America may, indeed, have a tough time prospering in a free-trade world where two such massive societies seem better positioned than we for an "energy showdown." We certainly must not underrate this real threat.

Let me summarize another way. Our allies—the EEC and Japan—are in worse shape than we are. But they are pushing ahead rapidly on the development of coal and nuclear energy.

It is imperative for all of us that oil supplies from the Mid-East remain available in the short run.

The controlled economies have fossil fuel reserves equal to ours—and lower labor costs. Even so, the USSR is pushing rapidly on fossil and nuclear.

The competition is very tough!

But the United States still has some things going for it that certainly *should* more than offset any opponent's apparent advantages. I can try to explain what I mean by repeating and amplifying a statement I made earlier in this paper: If the United States does not prosper in the years ahead, it will be because of reasons other than the higher cost of energy.

- It *could* happen that we will not prosper because we do not get government and industry working together for a common purpose on the international economic front and on the energy front—the way we do in defense when the chips are down.
- It *could* happen because we lose perspective of what we really do to ourselves when we try to over-regulate ourselves.
- It *could* happen that we will not prosper because we fail to

structure national policies in a way that will stimulate technological innovation—innovation to save and produce energy, and innovation to improve national productivity in a competitive world. Without a continuing flow of new high-technology products and promises, it is going to be really tough to be competitive with those countries that have ample energy and low labor costs.

- It *could* happen that we will not prosper because we fail to recognize that innovation requires more than new ideas; it requires *risk-taking investment* all along the line.
- It *could* happen because we forget where the funds for risk-taking investment come from.
- It *could* happen because we fail to remind ourselves that there must be rewards for those who have the courage to take the risks and who are skillful enough to win.
- It *could* happen that we will not prosper because we lose our national zest for life, our legacy of leadership—or because some Americans may believe the world owes them a living.

We are not owed anything; but nature and our forefathers have, together, given us much to do with for ourselves—if we will just do it.

Appendix
"Prospering" – How America Compares

		U.S.	EEC	Japan	USSR	China	Canada	Mexico
Population — Millions	1976	215	259	113	257	836	23	62
GNP — U.S. $s Billions	1976	1695	1442	574	717	307	183	66
GNP/Capita — U.S. $s	1977	7880	5575	5090	2800	370	7930	1060
Exports — U.S. $s Billions	1977	120	[a]382	81	45	8	43	4
Imports — U.S. $s Billions	1977	157	[a]390	71	41	7	42	5
Net Trade Balance — U.S. $s Billions	1977	−37	−8	+10	+4	+1	+1	−1
International Reserves — U.S. $s Billions	1977	19	96	23	12	?	5	2
Consumer Price Index	1977/1970	156	[b]188	204	?	?	165	?
Crude Steel Output — MM Tons	1977	113	126	102	147	25	14	?
Cement Output — MM Tons	1977	73	130	73	127	56	10	?
Grain Production — Kg/Capita	1977	1210	400	150	760	290	1730	16
Meat Production — Kg/Capita	1977	117	72	20	57	?	98	?
TV Receivers/Capita[c]	1974/1976	.62	.27	.23	.22	?	.37	?
Radios/Capita[d]	1973/1976	1.98	?	.70	.24	.04	.90	?
Telephones/Capita	1977	.72	.33	.43	.09	?	.59	?
Domestic Energy Production — Quads	1976	58	16	1	48	15	7	3
Energy Imports — Quads	1976	17	28	13	1	—	2	—
Net Energy Exports — Quads[e]	1976	−15	−22	−13	+7	+1	+1	—
Energy Exports — Quads	1976	2	7	—	8	1	3	—
Energy Use — Quads[f]	1976	73	38	14	41	14	6	3
Energy Use/Capita Q/100 Million	1976	34	15	12	16	2	26	5
Energy Use/GNP Q/$ Billion	1976	.043	.026	.024	.057	.046	.033	.045
Oil Production — Quads	1976	20	1	—	22	4	3.3	2
Oil Imports — Quads	1976	16	27	10	.3	—	1.6	—
Oil Exports — Quads	1976	—	4	—	6.3	—	1.3	—
Proved Oil and Gas Reserves — Quads	1976	420	260	2	1330	140	99	[g]67
Recoverable Coal Reserves — Quads	1976	5840	4090	175	4410	2570	175	30
Proved Fossil Fuel Reserves — Quads	1978	6260	4350	177	5740	2710	274	97
Estimated Uranium Reserves — Quads[h]	1977	256	40	—	?	?	90	—
Non-Electric Energy Consumption — Industry — Quads[i]	1976	19	10	4.6	?	?	1.4	?

Appendix
"Prospering" – How America Compares *(continued)*

		U.S.	EEC	Japan	USSR	China	Canada	Mexico
Non-Electric Energy Consumption — Transportation — Quads[i]	1976	18	5.6	1.8	?	?	1.6	?
Non-Electric Energy Consumption — RES, SMCL, Other — Quads[i]	1976	14	9.5	2.1	?	?	1.5	?
Price of Gasoline — U.S. $s/Gallon	1975	.50	1.15	1.05	?	?	.55	.33
Price of Electricity — Residential — U.S. $s KwH	1975	.04	.05	.05	?	?	.018	.05
Price of Electricity — Industrial — U.S. $s KwH	1975	.025	.03	.03	?	?	.012	.022
Electricity Produced — Quads	1976	7.2	3.8	1.8	3.8	.5	1.0	.2

[a]Exports and imports for EEC are gross trade estimates, which include intra-EEC trade.

[b]EEC estimate is very approximate. It is calculated as the GNP weighted average of consumer price indices for France, Italy, West Germany, and the United Kingdom.

[c]Data for U.S. and USSR are 1976; data for remaining countries are 1974.

[d]Data for U.S. and USSR are 1976; data for Canada, 1974; and data for China and Japan, 1973.

[e]Net energy exports equals exports minus imports.

[f]Refers to primary energy supplies available for transmission, refining, and conversion to/for final use. "Energy Use" is equal to domestic energy production plus imports minus exports.

[g]Oil and gas reserves for Mexico have substantially changed since 1976, and they are now estimated to be about 40 billion barrels or 225 quads.

[h]Uranium reserves are estimated as "reasonably assured resources" up to $30 per pound.

[i]Data source for U.S. is General Electric analysis; data for EEC, Japan, and Canada is from OECD energy balance sheet analysis.

Data collected with the special assistance of Lewin and Associates, Inc., 470 L'Enfant Plaza East, S.W., Suite 4100, Washington, D.C. 20024, from the following sources:

Sources: World Bank, *1978 World Bank Atlas,* Washington, D.C., 1978; United Nations, *World Energy Supplies 1972-76,* Statistical Papers Series J., No. 21, 1978; *The Oil and Gas Journal;* National Coal Association, *Coal Facts 1974-75;* International Energy Agency, *Energy Balance of OECD Countries 1974/1976,* Paris 1978; Central Intelligence Agency, Office of Economic Research, *Soviet Long Range Energy Forecasts,* September 1975, and *China: Energy Balance Projections, 1980 and 1985,* November 1977; Joel Darmstadter et al., *Energy in the World Economy,* Johns Hopkins Press 1971; Ford Foundation, *Nuclear Power Issues and Choices,* Ballinger, Cambridge 1977; and Workshop on Alternative Strategies, M.I.T., *Energy–Global Prospects 1985-2000,* Cambridge 1977.

Discussion

Chairman: Dr. Bueche, I first put you down as an optimist, one who really believes that our nation can prosper, at least relative to the rest of the world, despite higher prices of energy. Then you said that if we do not prosper it will not be because of energy, but because of an array of other considerations; and when you enumerated them, they seemed so certain and permanent and strong that they added up in my mind to a case that says we are not going to prosper. So I would like to ask again: are you really optimistic?

Bueche: Of course I am optimistic, and on two bases. First, I am already encouraged by some of the changes I see occurring today in capital formation and investment, and being able to make a little money. I think the income tax changes of 1978 were a step in the right direction, though they have not yet had their effect. Then too, the start of new businesses is on the upturn in spite of high interest rates and inflation.

My second reason for optimism is, history tells me that when the American people are faced with a job to do, a job they understand, a clear challenge, they respond positively and with vigor. I think if things get tight enough the American people and the Congress will be motivated to do the right thing. That is why I am optimistic.

Chairman: Your company, General Electric, has been one of the pioneers in development of nuclear technology. In view of the present concern over control and in the face of a virtual banning of nuclear reactors, are you optimistic that such a trend will be reversed?

Bueche: I think the anti-nuclear attitude will get worse before it gets better. A lot of people fear nuclear energy because they do not understand it. And when they talk about nuclear waste disposal (which *technically* is no great problem at all), they do not understand that either. They think it is going to seep out with all sorts of frightful effects. They are frightened.

But I maintain that if these things can be explained to the people, and if there is some leadership in Washington, both the fear and opposition can be handled. The United States *will* go ahead with nuclear power and coal and every other possible energy source—fusion, solar—and I am optimistic. It is going to happen.

From the floor: GE has had a very strong effect in fielding nuclear reactors. I wonder if you have looked at the question of putting them in places where the siting problems are less difficult and where perhaps they might have some side benefits?

Bueche: There is an idea on which some work has been done at General Electric, and which I find very exciting because not only does it help solve some of the problems we have been discussing, but it does something positive at the same time. The idea is to put twelve power stations—they need not be nuclear; coal would be fine—twelve thousand-megawatt power stations of the kind that are being built these days, along the St. Lawrence Seaway. GE has found that if twelve of these are sited along the Seaway, their waste heat would keep the locks open all year long, making Chicago an international seaport.

The New York Port Authority might protest, but this would change the entire economic complexion of the country, and it would also help reduce transportation costs. Cargoes would not have to be unloaded in New York or New Jersey and trucked or shipped on railroad cars all the way to the Midwest. The ocean-going ships could come up there all year long. It would be a tremendous economic development program, using heat that would have to be thrown away. And because the St. Lawrence has so much water going through it, the ecological effects on the river would be nil.

But a leader is needed, somebody who can get together with the Canadian government, which has to be our partner in this, and suggest it. If it were decided to have fuel reprocessing along that little string of the St. Lawrence locks, there would be minimal transportation for nuclear fuel storage or reprocessing. If it were decided to use coal, all the coal could be shipped in from the Midwest or the far West on barges and boats the year round to these coal-fired power plants along the St. Lawrence Seaway. It has great possibilities.

From the floor: According to your statistics, the United States stacks up relatively well energy-wise compared to our major competitors, Europe and Japan. Can we find a way to shift calls for energy sharing to the International Energy Agency in case of a shortfall in world supply?

Bueche: We have signed a sharing agreement.

From the floor: Should we "unsign" it?

Bueche: That is a touchy question. If we "unsign" it, I expect we will see a lot of changes in Europe and the Middle East—a lot of new alignments.

From the floor: You said that the United States will see worse times before it sees better times, and I gather you mean we have to have a period of shortages in order to unify the country to respond to the energy crisis.

Bueche: I think that is right. I do. Just pray that these days will be numbered, and not too difficult.

From the floor: The leadership will not come until there is a demand for it.

Bueche: That is not absolutely true, although it is much more likely that leadership will rise when leadership is demanded.

From the floor: As a technologist, would you comment on the future of solar energy, its economic viability vis-à-vis the other energy sources.

Bueche: I think the future for solar is very positive, and that it will not be very long before the costs of putting up good solar panels and all the plumbing and controls that go with them to heat water will be economic. In fact, I can give you some "ballpark" figures now. As of a year ago, if you lived in Los Angeles you could afford to pay $20 per square foot of good solar panels installed; that would cover all the pumps, pipes and storage tanks for hot water heating. And in Boston, you could afford to pay only $8 per square foot installed, because the weather is different and the oil prices were different there too. But now, as the cost of fuel goes up and the cost of solar panels and piping comes down (there is much technical work going on in this field and I am sure that prices will come down) then it should be economic.

As far as solar cells go—the direct generation of electricity from the sun—that will take somewhat longer to be economic across the board—that is, to replace central station power. But central station power need not be considered the ultimate goal. There are many specialized uses for solar energy, part-time uses, uses where it would be feasible to have it going only during the sunshine hours of the day, in which it would be quite beneficial and maybe economical on an incremental basis. It certainly will soon be economical (and perhaps already is) in some very remote sites to which it would be extremely costly and difficult to transport fossil fuels or charged batteries. But it will take quite a while longer to get down to broad application.

From the floor: Although the theme of the meeting here seems to be the development of energy sources through the private marketplace, most of the speakers have talked about some government role in developing new technology. What do you see as government's appropriate role in energy technology?

Bueche: Just as in everything else. I am like Abraham Lincoln on this. I believe the government should do for the people only what the people cannot do for themselves. Each case would have to be investigated separately. On the whole, I think government ought to clean up the regulations and get out of the way. *(Audience applause.)*

Chairman: As I recall, Dr. Bueche, you were heavily involved in the synthetic rubber program in World War II, in which government and private industry cooperated. Let us suppose we have a charismatic leader; people believe him and are willing to cooperate and do what he suggests. And let us suppose he suggests that a synthetic fuel program on a crash basis is absolutely essential to the United States, and he wants to use private capital and the free enterprise system and the know-how that exists in private industry to the fullest in setting up that program. Still, the government has to be heavily involved or it will never get off the ground. There must be a hybrid of free enterprise and government control.

He proposes to Congress and to us, the people, seeking support for Congressional action, that this massive program be set up in such a way that the government contracts to buy a given amount of the synthetic fuel—perhaps two or three million barrels a day—at the lowest price that competitive bids from industry will provide. There are, of course, certain environmental restrictions and quality requirements, and a long-term contract so well-bound that a future congress or president cannot change it.

Of course, private industry submit bids because there is a specific thing to bid on and it is possible to make calculations of return on investment. As a result the low bidder (with other requirements as to credibility and competence) wins and the program goes on. And even though the price is higher than the present OPEC price (and it undoubtedly would be), once the industry is created that would be the upper limit above which neither OPEC or anyone else could take energy prices. Now, what is wrong with that idea?

Bueche: What is wrong with it is that such a president would have to be a magician to convince the American people, and especially the American businessman, that those agreements would not be changed no matter how much Congress talked or debated or how much some special purpose group objected, no matter who was in the White House or what staff he had. It is not realistic. The synthetic rubber program of

World War II was, of course, in a time of crisis. Those who are old enough remember how serious it was. In three years, under the government program, production of synthetic rubber went from essentially zero to 700,000 long tons per year weight.

If I were to make a suggestion in the current energy situation, I would suggest that a program for synfuels, or whatever, be started out the same way, and that your proposal be worked in as a second stage. The first stage should be led by the government and the government pays for everything in that stage. The plants should be constructed by the government. There should be multiple plants for each of the major kinds of technology, and the individual projects should be in the hands of various companies, just as in the World War II synthetic rubber program where Firestone had one project, General Rubber had one, U.S. Rubber one, and so on. Each should be for a different kind of product. Once a month everyone should get together and tell about their progress and their problems in an open meeting.

Later on, if the plants work, they can be sold to industry at the proper price. But I think if it is necessary to do the job fast, it must be done that way.

Now, that would require a real leader, would it not? He could probably get industry to go along with such a plan, but unless there were a national crisis, Congress would crucify him. That is where we need leadership.

John P. Holdren

Energy, Economy, Environment: Interactions and Choices

The failure of the contemporary energy debate to converge on solutions results in part because the participants are not all talking about the same problem. It may be useful to devote the first part of this discussion, therefore, to an attempt to tie together in a single framework some seemingly disparate perspectives on what makes up the energy problem. That is most easily done by disaggregating the issues into three time frames relevant to the United States energy situation and four elements of the problem.

Perspectives on the Energy Problem

The time frames are historical (1940-1970), transition (1970-2010), and long-term future (post-2010); the four elements of the problem are technology, economics, environment, and politics.

John P. Holdren *is Professor of Energy and Resources at the University of California, Berkeley, and a consultant on energy to a wide variety of state, national, and international agencies. He is a member of the Department of Energy's Research Advisory Board, a consultant for the Solar Energy Research Institute, and serves on the Steering Committee of the Study of Nuclear and Alternative Energy Systems, being conducted jointly by the National Academy of Sciences and National Academy of Engineering.*

Historical Time Frame

In the historical time frame, the supply side of the U.S. energy situation was dominated by oil and natural gas, whose properties of easy transportability, storability, and adaptability to a wide variety of end-use needs lead me to label them the *convenience fuels*. Their share of the use of marketed energy forms in the United States grew from 43 percent in 1940 to 70 percent in 1970. The most important associated technological problems were those of exploration—how to locate the largest, most easily exploited deposits of these crucial substances.

On the economic side, energy in the historical period was cheap, not only because of the geologic "subsidy" represented by rich deposits of gas and oil concentrated over millennia by natural processes, but also because of subsidies both environmental and socioeconomic—the former often unconscious and the latter motivated by the pivotal role cheap energy was believed to play in driving economic growth. Energy and Gross National Product (GNP) grew roughly in step during most of this period, and this linkage was widely held to be not only causal but inflexible.

The environmental dimensions of energy were given relatively little attention during most of the historical period, with major initiatives aimed at raising the priority attached to environmental issues—the National Environmental Policy Act in 1969, Earth Day and the establishment of the Environmental Protection Agency in 1970—coming only at the end (indeed, for my purposes, helping *define* the end) of this period. Assessments of environmental effects, and measures applied to control them, were primitive. The great bulk of what little analytical attention went to these problems was devoted mainly to direct health effects of a few forms of air pollution and to conflicts between energy development and recreational values—a narrow view of environmental problems indeed.

From the standpoint of energy politics, it would be fair to say that "no-one was in charge." A few state and federal agencies had responsibilities related to energy, but there was little coordination and no coherent policy. Indeed, the most powerful and cohesive force in energy politics for most of the "historical" period was almost certainly the oil companies. And, except by them, international ramifications of energy issues were largely ignored.

Transition Time Frame

We find ourselves in the late 1970s in the first stages of a transition linking the era of cheap convenience fuels with a forthcoming era of essentially inexhaustible but much more expensive energy supplies. My choice of the year 2010 as the far horizon of the transition, while necessarily somewhat arbitrary, represents a time still close enough that the energy situation can be shaped substantially by today's knowledge, but far enough away not to be constrained entirely by the momentum of yesterday's decisions. (Experience suggests that completely new energy technologies require at least 40 years to proceed from conception to substantial impact on the national energy budget. Similarly, power plants designed in 1960 and entering commercial operation in 1970 will just be approaching the end of their useful lifetimes in the period 2000 to 2010.) The year 2010 is also a time when, according to the best presently available information about conventional oil and gas supplies, the United States will have to be well on the way toward weaning itself from its present heavy dependence on these fuels.

All the major energy studies of the past few years correctly characterize this transition period in the energy-society relationship as a time of major, interacting uncertainties—uncertainties, for example, about how much oil, gas, and uranium will still be found, about the nature and implications of the link between nuclear power and the spread of nuclear weapons, about the eventual economic cost of photovoltaic cells and other renewable-energy technologies.

In such a period, emphasis is bound to be placed on those energy sources that are *least* uncertain, that is, those we can most confidently count on, or, put differently, those least likely to be taken away from us by unpleasant surprises. For the United States, these sources are coal and conservation. I will call them the *reliable fuels*. (While some people may be unaccustomed to thinking of conservation—increased efficiency in energy conversion and use—as a "fuel," a Btu saved at the margin is fully equivalent to a Btu dug out of the ground.)

The dominant technological problems that will be associated with these sources are those of integration and matching with the pattern of energy end-uses that grew up around cheap oil and gas. Such problems include the conversion of coal to liquid and gaseous forms, and finding ways to improve the energy-using performance of existing buildings whose life expectancy (typically 75 years or more) precludes rapid replacement.

Economically, the transition period is being characterized by rising real costs of energy (that is, corrected for general inflation), arising partly from the depletion of the geologic subsidy and partly from the conscious phasing out of environmental and socioeconomic ones. Permitting the price of energy to approach the marginal cost to society of new supplies will lead to sounder decisions about how much energy to use, and the resulting substitution of ingenuity and nonfuel materials for energy will lead to a decoupling of energy growth from that of GNP. Such a period of decoupling would not be unique in U.S. economic history—the ratio of marketed energy use to real GNP fell by about 50 percent between 1920 and 1945—and the evidence is that the expected resumption of such a trend is now underway in response to the energy price rises of the mid-1970s. Although it is clear that the growth rate of energy use can fall without significant damage to the growth rate of GNP overall, there may be sectoral shifts in economic activity (from more energy-intensive to less energy-intensive sectors) that produce increased unemployment in some groups. This, and the differential impacts of rising energy prices on the poor, will be the most troublesome economic issues of the transition.

Environmentally, the transition period is characterized by growing recognition of the importance of energy's environmental dimensions, coupled with a broader definition of what these dimensions include: not simply the direct effects of pollution on public health, but the impacts of the full array of energy activities on "free" goods and services provided by biogeophysical processes, on occupational health, and on social and political conditions and relationships. The associated analytical problem that has emerged as central is how to characterize and quantify this diverse array of effects in a way that even permits meaningful comparisons between different technologies (the "apples and oranges problem"), not to mention formal integration into the benefit-cost analyses playing an increasing role in contemporary decision-making.

The politics of energy in the first part of the transition period has been characterized by increased government involvement and organization—a proliferation of bureaucracy and intensification of legislative interest—accompanied by the emergence of an array of interest groups and constituencies with conflicting priorities and policy proposals: consumer groups, trade unions, environmentalists, spokespersons for the poor and for particular geographic regions, and, of course, the energy companies. The question of whose interests and values will prevail is made the more complicated by the appearance of energy's international dimensions at center stage: balance-of-payments problems, the rich-poor gap, links between nuclear power and the proliferation of nuclear weapons, and so on.

Long-Term Future.

Energy supply for the long term will be based, by definition, on the inexhaustible fuels: solar energy in its many manifestations (which include wind, hydropower, ocean thermal gradients, and biomass); nuclear fission using breeder reactors; nuclear fusion; and possibly some forms of geothermal energy. In all of these cases, the fuel is free or nearly so, but the technology required to convert the energy into the forms desired is complicated, difficult, materials-intensive, or all of these.

Economically, then, none of the long-term energy sources can be expected to be cheap. The long term will be an era of permanently expensive energy, by comparison with the three-dollar-per-barrel oil and five-dollar-per-ton coal of a few years ago. But the most difficult economic issue connected with energy in the long term will not be energy's high cost alone, but the consequences of a rational society's inevitable eventual decision to limit the rate at which energy is used. This decision will come when, despite the best technologies, the marginal social costs (economic plus environmental plus sociopolitical) of increasing the rate of energy use begin to exceed the marginal benefits. Because substitution of other inputs cannot reduce energy's role in economic production indefinitely, stabilizing energy use eventually will mean facing up to a limit to economic growth itself.

The inevitability of the existence of a point where the marginal social costs of further energy growth exceed the marginal benefits follows mainly from a property of the environmental part of those costs: beyond some level of energy use, which depends on the technology, environmental costs rise very steeply with further growth. This central and evidently inescapable role of environmental characteristics in determining the maximum rate at which a rational society will use energy—and hence, ultimately, how much energy-derived well-being can be delivered—means that these characteristics probably will govern the choice of energy technologies in the long term. Even substantial differences in economic costs, in other words, will be likely to be outweighed in importance by any relative environmental advantage of one mix of sources compared to another.

The political dimension of energy in the long term is more difficult to predict. The interaction of patterns of energy supply and use with the social and political characteristics of the society surely will be more generally and explicitly recognized than in the case today, and one would expect more success than today in integrating energy choices with social goals. How that integration will affect any given aspect of energy politics, such as degree of centralization in energy management and control, remains an open question.

The United States in Transition

The foregoing set of perspectives suggests some propositions about the U.S. energy situation at present—the early part of what I have called "transition"—that deserve emphasis.

First, the nation is not running out of energy in any absolute sense, but only out of the one-time windfall of cheap and convenient oil and gas. The principal immediately available alternatives—offshore, Alaskan, or imported oil and gas, coal, nuclear power, solar heat—are more expensive, less convenient, more threatening environmentally or politically, or all of these. What the present political trauma over short-term energy choices reflects is not an inability to supply more energy at some cost, but a growing perception in some constituencies that the cost is too high.

Second, expensive energy is not temporary; energy will never be cheap again. It may be a modest consolation in this situation to note that the cheapness was an illusion all along—an artifact of geologic, governmental, and environmental subsidies. There are people in the consumer movement and the environmental movement (with whom I am in sympathy on many other issues) who want to keep energy artificially cheap. On this point I believe they are misguided. There is no surer prescription for postponing the needed transition toward sustainable energy sources and increased efficiency of energy end use. At the same time, acceptance of higher prices does justify concern with two related questions: Who gets the money? How can we ameliorate the burdens of high energy costs on the poor?

Third, the growing role of environmental issues in shaping energy choices is not a temporary aberration. Rather, it is part of a natural and inevitable transition to a decision-making framework in which maximizing well-being, not energy for its own sake, is the goal, and in which, therefore, environmental and economic services and costs are considered together.

The Conservation Option

In the present situation, it is becoming clear that the easiest, fastest, cheapest, safest, cleanest "new" energy source is energy conservation:

- It is easier and faster to increase availability of natural gas by insulating attics than by drilling for it in Alaska or offshore.
- It is cheaper and safer to save gasoline in buses and more efficient automobiles than to pay for the extra consumption of gas guzzlers with expensive imports subject to interruption.
- It is safer and cleaner to make electricity available by reducing the amounts used by refrigerators and air conditioners than to build additional nuclear and coal-burning power plants to run inefficient ones.

It should be clear from these examples that by "conservation" I do *not* mean curtailment, belt-tightening, or sacrifice. Rather, "conservation" should be understood as using less energy to produce the same goods and services, or by shifting to a less energy-intensive mix of goods and services that produces the same perceived well-being. It means, in other words, increased efficiency of energy use—the logical response to rising energy costs.

Specific possibilities for energy conservation can be divided into four categories, as follows:

1. *Retrofit* of existing structures and equipment to improve their efficiencies (for example, addition of heat recuperators to exhausts from industrial processes where heat is used, and the addition of insulation to walls and ductwork in commercial and residential buildings).

2. *Replacement* of existing energy-using structures and equipment with more efficient (energy frugal) designs, when the economics makes if favorable to do so (for example, introduction of energy-efficient cars, appliances, and buildings).

3. *Research and innovation* to develop entirely new technologies that deliver equivalent or greater services compared to existing approaches, while using little energy. (Past or pending products of such research include pocket-size electronic calculators to replace typewriter-size mechanical ones, tiny microfiche to store information with far less bulk and resource use than paper, and computerized banking to minimize physical transport of "hard-copy" material in financial transactions.)

4. *Reasonable restraint* in avoiding forms of behaviors that squander energy without contributing to real well-being (thermostat settings that overheat buildings in winter and overcool them in summer, compared to requirements for comfort and health; highway speeds that not only guzzle gas but maim and kill; failure to turn off lights and close drapes, resulting in illuminating empty space and heating the great outdoors).

All these approaches involve substituting something for energy—sometimes labor, sometimes nonfuel materials, sometimes capital, sometimes only ingenuity. But none represents sacrifice, because the substitutions are sensible—they make people better off, not worse off.

How much energy can be saved? No-one can say exactly, in part because available resources of technological ingenuity have barely begun to be applied to the task. But the most thorough investigation done to date of the situation in the United States indicates that cost-effective application of existing technology could cut energy requirements per dollar of GNP *two-fold* by the year 2010, assuming energy prices continue to rise. That is, one could have twice the real GNP in 2010 as in 1975, with no increase in energy use.

Energy Supplies beyond Conservation

Conservation must be the cornerstone of energy policy for the transition, but clearly everything cannot be conserved. Even if the total rate of energy use in the United States did not increase above its present level, choosing the best mix of supply options to maintain that level, as the contributions of conventional oil and gas decline, would present challenging problems. Furthermore, less developed countries typically have much less fat in their present energy budgets than does the United States, and they have a much more compelling need to achieve a high rate of economic growth. Thus their energy requirements can be expected to grow much more rapidly than ours. Because, however, the richest industrial nations necessarily do most of the world's research and development related to energy supplies, the United States has a clear responsibility to keep poor country needs in view along with its own.

The menu of energy resources and energy-conversion technologies from which this country's and the world's energy choices for the future will be made is long—conventional and unconventional oil and gas; oil shale; coal, conventionally burned or for synfuels plants; nuclear fission in conventional or breeder reactors; nuclear fushion; geothermal energy in half a dozen forms; and renewables (mainly solar heat, biomass, wind, hydropower, ocean heat, and solar electricity). Even that is only a partial list. Some people say the situation requires going full speed ahead on all of them, conservation notwithstanding. I reject that view. It is true that diversity is a form of insurance against an uncertain future but, as in all insurance policies, diversity is expensive. We simply cannot afford to develop everything.

Elements of Choice

How, then, should we choose? I believe the most important criteria are these: economy, expandability, versatility, environmental tolerability, and certainty. The first one, economy, is clear enough. Expandability means production has the potential to grow with needs, if not to expand the availability of energy in total, then to replace sources whose production is declining. Versatility means the source can be matched to a variety of kinds of end-use, at a variety of scales from big to small, in a variety of geographic and cultural contexts. Environmental tolerability means we can live with it, or, more sensibly, that the environmental and social costs of its use do not outweigh the economic benefits. Certainty means we can count on it: we know it works; we control its availability; our uncertainties about its other characteristics are tolerable.

Using these criteria, and homing in—in the interests of time—on those energy options that could make a major difference relatively soon if called upon, permits a rather drastic reduction of the menu. Oil and gas, for example, are part of the problem more than they are part of the solution. Excessive dependence on foreign oil damages our balance of payments, threatens to compromise our foreign policy, leaves us vulnerable to sudden embargoes, and places us in competition with industrialized allies and developing countries who need the oil more because their options are fewer. The amount of domestic oil and gas that will ever be extracted is still open to debate, and breakthroughs in advanced recovery or unconventional gas could buy some unexpected time; but the plain fact is that the only certainty is a decline in the fractional contribution of oil and gas to the national energy supply. Geothermal energy is locally interesting (in California, for example), but cannot play a major role in the national picture before sometime in the next century, if then. The breeder reactor is clouded by uncertainty about technology, economics, safety, and the vulnerability of the plutonium fuel cycle to terrorism and proliferation. It possibly might emerge from all this as a viable energy source for the long term, but it cannot now be counted on and it cannot be important before 2010. Fusion is fascinating but still far away; it is not likely that it can be important much before 2020.

Most Promising Energy Options

The main contenders that remain are the solid fossil fuels (that is, coal and oil shale), conventional nuclear reactors (that is, nonbreeders, not recycling plutonium), and various of the renewables. How do these compare? All could be expanded greatly over the next few decades, at least as far as resource availability is concerned. (Uncertainty about uranium availability poses the possibility of a lid on light-water reactor capacity between 2000 and 2010.) With respect to versatility, nuclear power has the least of the group, producing only electricity and even that only in very large blocks. The renewables as a group have the greatest versatility, including especially their applicability in developing countries. Coal, oil shale, and one of the renewables (biomass) can all help plug the fluid-fuel gap, but only at the expense of considerable and costly technological complexity. None of these sources, in fact, offers any reason to suppose I was wrong to say energy will never again be cheap. Coal and nuclear power are roughly competitive economically for electricity generation under present conditions, leaving aside for a moment nuclear's vulnerability to safety-related shutdowns. Coal and oil shale may be roughly competitive for fluid-fuel production. The unconventional renewables (that is, all but hydropower), which are the ones with the greatest potential for expansion, have the reputation of being generally more expensive than the cheapest conventional alternatives for the same end uses, but I believe the pace of developments in this area is now such as to make it difficult to make a sensible statement that will not soon be overtaken by events.

In this situation, one might be content to let the economic marketplace make the choices for us, except for one enormous "catch": the environmental and social costs that escape the economic balance sheets of energy sellers and buyers. I argued earlier that these external costs increasingly are and ought to be central to our energy choices. What can be said about them?

Environmental Liabilities and Their Implications

First, no energy option known or imagined—including energy conservation, solar energy, and fusion—is completely free of environmental liabilities. But these liabilities vary greatly in severity, particularly when one accounts not only for uncorrected impact but for the possibility of amelioration by improvements in technology or management. Among the supply options on which I have focused here—coal and oil shale, conventional nuclear fission, and unconventional renewables—the greatest direct damage to human life and health in routine operation almost certainly is done by the fossil fuels. The main villains are respiratory illness from emissions of sulfur oxides and particles, coal-mining accidents, and (would you believe?) auto collisions with coal trains at grade crossings. What is too rarely emphasized is that all these hazards can be very substantially reduced by known methods (for example, in the last instance, putting in overpasses and underpasses at grade crossings). Also too rarely mentioned is that the shocking health casualties often attributed to coal in public discussions—100 excess deaths per large power plant per year—are at the high end of a range of scientific uncertainty that extends down to zero on the low end.

It may well be that the most serious environmental costs of coal and oil shale are not the direct effects on health but the damages to ecosystems and hence valuable environmental services. These damages may arise from consumptive water use in the semi-arid regions where much coal and oil shale are found, from acid rain formed from sulfur oxides and nitrogen oxides released in combustion, and from climate change induced by the build-up of fossil fuel-produced carbon dioxide in the atmosphere. For the last problem, no plausible cure has been proposed short of limiting the use of fossil fuels.

Uncertainties of Nuclear Fission

The most serious environmental problems with nuclear power, in my view, are its links to the proliferation of nuclear weapons among nations, and, where plutonium recycle is practiced, the possibility that subnational groups will steal this materal and construct their own nuclear bombs. Neither of these issues is susceptible to persuasive technical analysis, and the uncertainty about what to expect is enormous. The proper question with respect to the influence of these problems on the acceptability of nuclear power, in my view, is this: Is this uncertainty *itself* a reasonable trade for the benefits of nuclear power?

I believe that the risk of reactor accidents is a somewhat less threatening problem than the first two, because the potential effectiveness of strictly technological safeguards is greater. Nevertheless, as the Three Mile Island incident emphasized, human frailties play important roles—these frailties, of course, could in the future include malevolence as well as carelessness and incompetence. Neither operating experience nor the presently available combination of theoretical and experimental studies is adequate to pin down the risk of nuclear reactor accidents (accident probability times consequences, added up for all possible malfunctions) to a narrow range. At the high end of the range of values that is conceivable, nuclear power would be a considerably greater direct hazard to human life and health than getting the same amount of electricity from properly sited and managed coal plants. Of course, nuclear power produces no carbon dioxide—but then, coal has no links to nuclear bombs. You will note the difficulty here that professional comparers of energy risks call "the apple and oranges problem."

The Renewables—Striking a Balance

Renewable energy sources vary as widely in their environmental effects as in their technological characteristics. Many make heavy demands on land, with potential damage to ecosystems. Some make heavy demands on materials to construct the equipment, which leads to some emissions and occupational injuries in materials acquisition and fabrication of components. Large-scale biomass plantations, if heavily irrigated, fertilized, and pesticided, could be quite destructive ecologically. All available evidence indicates, however, that reasonable care in selecting which renewable energy technologies are used could make available large amounts of energy with considerably less troublesome environmental costs than either the abundant fossil fuels or nuclear fission. This means not only that the environmental costs tend to be smaller per unit of energy delivered, but that the biggest of these costs tend either to be relatively easy to bring into the economic balance sheets (e.g., compensation for occupational injuries being incorporated in the price of the energy) or at least to be borne by the beneficiaries of the energy use (e.g., the owner of a rooftop collector bearing the risk of damage if it leaks). This last set of characteristics is in sharp contrast to those of the fossil fuels and nuclear fission, many of the most serious environmental costs of which are exported in space and time far beyond the users of the energy. If such exported costs are big, of course, the circumstance that the energy users neither bear them nor have to compensate the victims is both unjust and inefficient—the latter because users who do not pay the full costs of their consumption will not make socially efficient decisions about how much or what kind of energy to use.

Conclusions

I believe—and hope—that these very significant advantages of many renewable energy technologies on the environmental side of the energy/well-being relation—by comparison with fossil fuels and nuclear fission—will lead to an accelerated displacement of the latter by the former in our energy mix, beyond what would be brought about by market forces alone. I cannot emphasize enough, of course, how much the practical potential of this hope hinges on whether the growth rate of energy use as a whole in this country in the next few decades is high or low. If it somehow were to remain as high as the average of the last few decades, the result would certainly be attempts to deploy virtually every possible source on a crash basis. That is, there would have to be a major expansion of nuclear power, uncertainties be damned; a rapid expansion of coal, the dirty and disruptive technologies along with the relatively clean; and large oil imports, with all the costs those entail. And in this situation, the *relative* role of the environmentally more benign renewables would be small. There is much reason to believe that in terms of real impacts on well-being, this future would amount to too much energy, too soon.

Low growth, by contrast, offers the opportunity to shape the energy future more sensibly. It lets us choose the most suitable technologies, rather than having everything forced down our throat. It buys the luxury of proceeding slowly on nuclear power, and gives us time to bring to bear the ways we know exist to mine coal more safely and burn it more cleanly. It holds out the hope of keeping us below the threshold of a carbon-dioxide-induced climatic debacle. And low growth gives renewables a chance to make a real dent.

Nor need a low-growth energy future be synonymous with too little energy, too late, because the conservation that brings about the low-growth future will not be curtailment—doing without—but rather will be increased efficiency—doing better, milking more prosperity out of each gallon of oil, each pound of coal, each kilowatt-hour of electricity. It is this rational approach to more efficient energy use, first and foremost, that provides us the opportunity to steer a judicious course between the twin perils of too little energy and too much.

Discussion

From the floor: When do you project the possibility of allowing more decentralized energy sources in order to have those who benefit pay the cost? Would that not have a kind of dropout effect, because without the participation of groups that do not benefit it may not be possible to support the large enterprise necessary for central power, which is essential for industry in many cases?

Holdren: Much more study is needed on how one would integrate the use of decentralized energy resources with a continuing degree of centralization which, as you suggest, will be required for the kind of society we have today. We cannot go all the way toward a decentralized system, but what the boundary between the two would be remains to be seen. Some interesting studies of that are starting to emerge. I do not see it as an insurmountable problem at this point but it needs more attention.

From the floor: Considering the difficulty of separating environmental costs from benefits, cannot a case be made for giving a free market a stronger role? What do you think of a bidding approach by which communities that are being considered as sites for a power plant tell how much they expect to be compensated before they are willing to bear the localized environmental costs of having the plant?

Holdren: The bidding concept was raised at the Environmental Quality Lab at Cal Tech in '72 and '73, and was pursued quite vigorously at that time by social scientists, economists, technologists interested in these questions. We presently have a study underway in my program at Berkeley of bidding approaches to internalizing environmental costs. The market can certainly do a lot more than it has been allowed to do so far. However, there will remain some resistant and intractable problems where the degree of export of the costs both in space and time is so great that one may not be able to find a way to bridge the gap.

From the floor: You are clearly well acquainted with the environmental community which has considered many of the factors you have discussed. What, in your judgment, is the extent to which the environmental groups, broadly speaking, have reached their conclusions and rejected technology, as they see the future somewhat as the Luddites did back in the early days of the Industrial Revolution?

Holdren: That would not be a fair characterization of most of the environmental movement. Of course there are strange bedfellows in all areas of the energy debate; all of us sometimes take the same positions for different reasons than other people take them. I was trained as a technologist and believe the key to many of these problems lies in the enlightened use of technology and innovation. And I believe most people in what I call the rational environmental community agree with me—that their position is not anti-technology, but in favor of an enlightened use of the best technologies that can be brought to bear. This includes enormous opportunities in renewable energy sources, in energy conservation, in using fossil fuels more cleanly, and possibly in making the fission option one that would be viable. I am an agnostic on whether fission will be needed or desirable as a long-term energy source for the world. My present perception is that its problems have such a big component of the intractable human factor in them that it is questionable they can be resolved. So I would prefer to put a lot of technological investment in alternatives.

From the floor: Do you see a way in which alternative energy options might be rated on the basis of social equity?

Holdren: It would probably be premature to do so at this point, but I will give some preliminary impressions. The renewable sources would seem to have some intrinsic advantages in this respect. (It remains to be seen what disadvantages they may have.) The first intrinsic advantage is that they rely upon resources that are widely distributed. For example, in parts of the world where poverty is most prevalent today, the natural resources include typically a lot of sunlight for conditions capable of growing biomass in larger amounts. Second, in many cases, these technologies can be harnessed at very small scale as economically as at large scale. The difficulties of technologies like nuclear fission and nuclear fusion and, to a lesser extent, coal, from the point of view of the dispersed and generally small-scale energy needs of the developing countries, is that these more sophisticated industrial nation technologies are usually economic only in unit sizes of 500 megawatts or more. Until recently, people thought fusion might be economic only in sizes of, say, two gigawatts or more; recently that impression has changed. It now appears likely that fusion will be economic in the few hundred megawatt zone, but that is a big problem for developing countries. Third, again in terms of equity, is this matter of how the costs and benefits are distributed.

From the floor: If the model of human aspiration is total liberation from time and space—intercontinental travel over the weekend, picnics on the moon, the Buck Rogers-Star Trek scenario—in essence per capita energy requirements could be infinite. Do you accept this model as valid?

Holdren: I accept the validity of visions of a long-term future that includes all kinds of things we can scarcely talk intelligently about today. Even thinking fifty or a hundred years ahead, some people talk about a world population of twenty billion persons and average use of twenty kilowatts-per-person, which means 400 terawatts as civilization's energy use compared to eight today. That is not what we are discussing here and now. We are talking about what a mathematician would call an initial value problem; that is, we have a starting condition, and the real difficulty is how we get from where we are to any sort of long-term equilibrium future. "Where we are" is in a world of four billion people, one third of whom use 85 percent of the available energy with the remaining 15 percent divided up among two-thirds of the earth's people. It is a world of crushing poverty, with enormous threats of instability and nuclear war. Somehow it is necessary to come to grips with all these dynamic *transition* problems, in order to have the luxury of even thinking about the sort of world you have mentioned.

W. Kenneth Davis

Abundant Energy, Acceptable Price— A Realistic Goal

The early 1970s were marked by a series of truly significant events that fundamentally and permanently changed United States energy prospects, but the consequences seem to have been accepted with reluctance, if at all, by the public and our legislators and administrators. The important events included:

1. The peaking of U.S. oil production in 1970;
2. The peaking of domestic natural gas production in 1973;
3. The OAPEC (Organization of Arabian Petroleum Exporting Countries) oil embargo in 1973-74, which, while temporarily reducing U.S. oil supply by only about 5 percent, seriously disrupted the economy and clearly demonstrated the inherent risks of large-scale energy imports;
4. Massive increases in OPEC oil prices at the time of the embargo, with subsequent price increases for both oil and other competitive fuels;
5. A rising tide of public sentiment to preserve the environment at all costs, which has effectively precluded the timely development and

W. Kenneth Davis *is Vice President of the Bechtel Power Corporation, which designs and constructs thermal power stations. He was Director of Reactor Development for the Atomic Energy Commission from 1955 to 1958. He is past President and honorary director of the Atomic Industrial Forum, and has served on numerous national committees having to do with alternative energy sources, including radioactive waste management and other nuclear matters. A chemical engineer, he is the newly-elected Vice President of the American Institute of Chemical Engineers and the Vice President of the National Academy of Engineering.*

deployment of almost all new energy sources and facilities in the United States.

Our national impotence to deal effectively with this situation has seriously decreased U.S. credibility and power in the world community and has increased the risks to our national economic health and security far beyond any reasonable and acceptable limits. It is particularly disturbing that despite rising prices and an unfavorable long-term outlook for domestic production, U.S. oil consumption has increased from about 17 million barrels per day (BPD) in 1973 to about 19 million BPD in 1978 and 20 million BPD a year later. Oil imports have grown from 37 percent of total supply in 1973 to about 44 percent today, despite the added production from Alaska. In 1973 about 8 percent of this oil came from OAPEC countries; today, more than 16 percent comes from the Arab countries.

Whereas the possibility that future oil supplies may be inadequate to meet demand is a source of grave concern to the industrialized countries (it is generally believed that world-wide oil production will "peak" around 1995, give or take a very few years), it is a pervasive source of dismay and consternation to the oil-short, developing countries whose energy consumption is increasing dramatically as they struggle to raise their standard of living.

The radical change in U.S. energy prospects has led to two disparate and equally unrealistic lines of argument, which have confused the public as well as our elected and appointed representatives. The first argument is that all we have to do is "conserve" and there will be enough energy for everyone. The "Good News About Energy" as reported in a 1979 publication of that title by the Council on Environmental Quality (CEQ) is that energy is not needed. While it is true that the rate of energy demand growth has been and can be further slowed by higher energy prices, by intelligent avoidance of waste and unnecessary use, and by increasing use of energy-efficient equipment, structures, etc., conservation alone will not solve the problem, or come close to it. The "final" conservation solution, perhaps the result of self-fulfilling prophecy, is absence of supply when needed, shortages, rationing, etc., with catastrophic effects on our economy, standard of living, and perhaps our national existence.

The other line of specious argument is that new sources of energy, particularly solar energy, are "almost available" and will be cheap, clean, and environmentally benign. While there may be marginal economic applications, particularly for solar heating, any realistic and responsible evaluation of the time required for large-scale use of solar energy for a significant part of our energy needs is well into the 21st century, assuming that the outstanding technical and economic problems can indeed be solved. Other instant "magic" solutions such as fusion, hot-rock geothermal, wind, and OTEC, are equally unrealistic for at least the next several decades. Irresponsible statements to the contrary have deluded the American public into believing that there are impending new technological solutions so there is no need to worry (except perhaps that the oil companies and "big business" may be preventing the introduction of such technologies) and particularly no need to make difficult decisions now on a national scale.

The simple fact of the matter, which is the subject of this paper, is that the United States has adequate resources to meet its energy needs for the next several decades—and, in fact, is uniquely blessed with such resources. It is our contention that if we could "get ourselves organized" in an intelligent fashion, all U.S. energy needs for the next several decades to ensure a healthy economy and an increasing standard of living could be provided at an acceptable level of risk to public health and safety. Further, we conclude that this energy could be delivered at an average real price (constant value dollar) only about 50 percent greater than prices today, increasing the annual expenditure for energy from about 12 percent of GNP in 1978 to 15 percent in the year 2000. This would seem an acceptable price for maintaining a high standard of living and assuring national security, at least as affected by energy concerns.

Evolution of the Present Energy Situation

The growth of the U.S. economy from the turn of the century through the 1960s was fostered by the nation's ability effectively to utilize its abundant sources of low-cost domestic energy resources. As the demand for energy increased, coal replaced wood as the principal energy source; coal was subsequently displaced by a progressive shift to oil and natural gas. Substantial improvements in the technology for recovery, transport, and utilization of these resources was achieved as well, resulting in a gradual decline in real energy prices (Figure 1), encouraging increased consumption.

Although the cost of exploration, development, and recovery of new domestic supplies of oil and gas inevitably began to increase, the ready availability of low-cost foreign sources of petroleum as well as the control of natural gas prices by the federal government at levels

Fig. 1. Primary Fuel and Electricity Price Trends

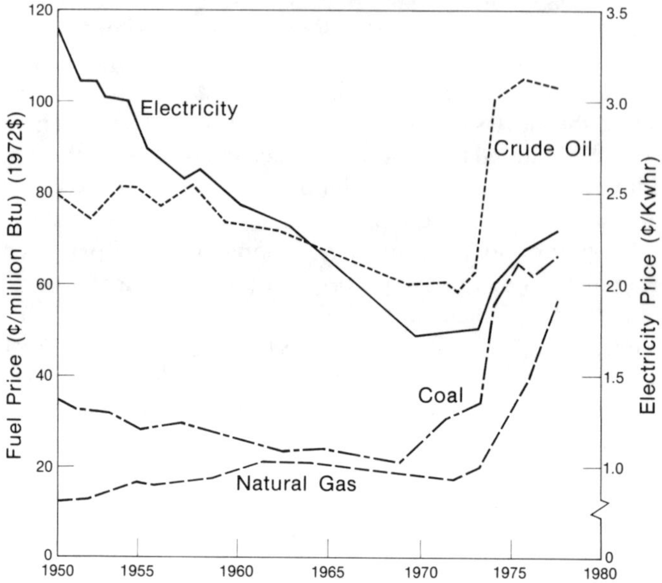

Source: Bechtel Power Corporation, May 1979

substantially below their replacement cost delayed the onset of higher prices to consumers. This provided a sense of euphoria that the era of low-cost energy could be sustained indefinitely, and by 1972, the U.S. had become dependent on oil and natural gas for 78 percent of its total energy supply.

The oil embargo of 1973/74 brought about a dramatic reversal of the downward trend in energy prices. The situation was compounded by a number of other factors that acted to drive up energy prices, including:

- Partial decontrol of natural gas prices which had previously been held at artificially low levels;
- Substantial increases in both capital and operating costs for energy supply facilities, reflecting in part the increasingly strict requirements for environmental protection and plant safety;
- Abnormally high costs for financing energy projects resulting from high interest rates associated with a high rate of inflation;
- Substantial lengthening of schedules for energy supply projects due to escalating licensing requirements and extensive delays associated with obtaining site approvals;

- Higher fuel costs due to
  - A long-term decline in resource discovery rate per unit of exploratory effort,
  - Increased mine safety requirements,
  - Inability to burn lower-cost, high-sulfur fuels due to air quality regulations,
  - A temporary uncoupling in some instances of the normal relationship between fuel prices and their production cost,
  - Increasing dependence on high-cost imported oil, and
- A host of political/institutional problems preventing increased utilization of abundant coal and uranium resources.

Matching Energy Consumption with Relative Abundance of Energy Resources

Despite the severe economic and national security impacts resulting from the oil embargo, the United States has accomplished little in the post-embargo period to reduce its overdependence on oil and gas.

Table 1 illustrates the pattern of U.S. energy consumption versus the relative abundance of domestic energy resources. A comparison of the

Table 1. United States Energy Consumption, 1978

	Relative Energy Consumption				Relative Resource* Abundance (Percentage)
	Residential & Commercial	Industrial	Transportation	Total	
Coal	9.4%	10.3%	0.1%	19.8%	40.6%
Natural Gas	13.1	12.4	0.7	26.2 }76.1	0.2
Petroleum	13.0	11.4	25.5	49.9 }	0.2
Nuclear	2.4	1.6	0.1	4.1	52.3
Oil Shale	0	0	0	0	6.7
	37.9%	35.7%	26.4%	100.0%	100.0%

Sources: Monthly Energy Review, U.S. Department of Energy, Energy Information Administration, March 1979; "U.S. Energy Supply Prospects to 2010," the Report of the Supply and Delivery Panel to the Committee on Nuclear and Alternative Energy Systems, National Research Council of the National Academy of Sciences, 1979.

*Mineral resources only; excludes hydroelectric, geothermal, and solar energy. Nuclear resources assume ultimate extraction of approximately 60 percent of the energy content of uranium by means of breeder reactors.

last two columns makes it clear that real progress toward energy independence can be achieved only by increased use of coal, uranium, and oil shale, and reduced dependence on petroleum. These solid fuels and the products from their conversion to synthetic fuels and electricity must ultimately be substituted for natural gas as well where practical, although this need is somewhat less urgent since natural gas imports at present are negligible and are likely to increase relatively slowly.

The most difficult problem in reducing petroleum demand is that more than half of all oil consumed is used for transportation. Because no other energy resources currently available are directly substitutable for gasoline, diesel, and jet fuel, prompt development and widescale deployment of facilities to convert coal and oil shale to synthetic liquid fuels are essential. Electric vehicles may ultimately permit some substitution of electricity for gasoline, but a meaningful contribution is unlikely before the turn of the century.

A large fraction of petroleum as well as natural gas consumption in the residential, commercial, and industrial sectors is used for space heating and water heating. Insulation of existing buildings and energy-efficient design of new structures could slowly but significantly reduce the total energy demand for space heating at a relatively reasonable cost. In locations where the climate is mild, solar heating may prove promising, particularly in new buildings. Electrically driven heat pumps and electric resistance heating can provide a substantial portion of heating needs. Direct burning of coal for heating could also substitute for oil and gas if judged environmentally acceptable, but very stringent regulations on emissions make extensive utilization of coal burning in small-scale applications very doubtful.

Synthetic liquids and gases produced from domestic oil shale and coal as well as from Canadian tar and oil sands could ultimately displace a significant portion of the demand for conventional petroleum and natural gas for space and water heating, especially in existing buildings where the cost of retrofitting to utilize electric or solar-based alternatives is economically impractical. Synthetic fuels could also serve as feedstocks for petrochemical production.

A continuation of the long-term trend toward increasing electrification appears highly probable. Despite the higher price of electricity relative to other energy forms, its cleanliness and convenience have made it an increasingly desirable form of energy. Also, nuclear energy as well as most other "new" energy sources can be most effectively utilized in the form of electricity.

Assuming a growth in GNP of 3 percent per year and a 1 percent escalation of real electricity prices, and using the relationship presented in Appendix B, we estimate that the average rate of growth of electricity demand in the period 1979 to 2000 will be 4.8 percent per year. Whereas geothermal power and hydroelectric power could provide a small portion of the additional generation needed, the contribution of these sources, while significant, could offset only a small fraction of the total demand. The bulk of new capacity would have to be met by substantial future growth in both coal and nuclear power plants.

One Possible Approach to Stabilizing the Energy Situation

In order to assess the future trends in energy prices, let us postulate what the energy situation *might* be in the year 2000, using the following somewhat arbitrary set of ground rules for future energy demand:

- Average growth in real GNP of 3.0 percent per year from 1979 to 2000;
- Growth in total energy consumption during this period of 2.5 percent per year, reflecting continuing substantial energy savings from conservation;
- A continuation in the long-term trend toward increased electrification, with energy used for electricity generation rising from slightly over 30 percent of total energy consumption in 1978 to 48 percent* in the year 2000;
- Reduced dependence on energy imports, especially those from outside North America; and
- High priority assigned to increased use of coal, uranium, and oil shale, and reduced use of oil and natural gas.

Specific actions assumed to meet this demand while minimizing the increase in energy prices include:

- Making a national commitment to the development, demonstration, and deployment of technologies for producing synthetic liquid fuels from coal and oil shale;
- Implementing a similar though less urgent program to produce synthetic natural gas from coal;

*This represents a growth rate of about 4.8 percent per year for electricity, as contrasted with only 1.1 percent per year for non-electric energy consumption, and 2.5 percent per year for total energy consumption.

- Deploying where economic both solar energy and electric heat pumps for space and water heating as a replacement for hydrocarbon fuels;
- Facilitating the use of lower-priced Western coal by permitting mining on federal lands, removal of existing obstacles to coal transport by slurry pipeline, and reducing requirements for SO_2 removal from low sulfur coal;
- Taking steps to facilitate meeting the bulk of future growth in demand for electric power with coal-fired and nuclear power plants;
- Initiation of a realistic North American Energy Policy which would reduce the risk of interruption of imported oil and natural gas supplies by increasing the fraction of these fuels of North American origin through the development of strong, mutually advantageous economic alliances with Mexico and Canada;
- Re-assessment of U.S. environmental policy and modification of it as necessary to achieve an appropriate balance between the relative costs and benefits of environmental protection;
- Simplification of the licensing process for all types of energy facilities;
- Reformation of utility ratemaking procedures to permit justifiable rate changes without unreasonable delays, thus restoring the ability of utilities to finance new energy projects at competitive rates;
- Elimination of price controls on all fuels to encourage an expansion of supply and a reduction of demand; and
- Continuation of a high level of research, development, and demonstration of energy supply alternatives to meet the nation's needs in the 21st century, with particular emphasis on solar photovoltaics, liquid metal fast breeder reactors (including necessary nuclear fuel reprocessing technology), and geothermal technology.

Taking action to implement these recommendations could lead, in the year 2000, to a pattern of primary fuel consumption similar to that shown in Table 2. This distribution is neither claimed to represent an optimum strategy nor to be one that could be easily accomplished, but if implemented, it would be a major step toward insuring a secure energy future for many years.

Our assumption that U.S. coal consumption will be 22.0 MBOE/day (2064 million short tons/year) in the year 2000 is 20 percent or so below

Table 2. Energy Sources

	MMBOE/Day*		% of Total	
	1978	2000	1978	2000
Coal	**6.7**	**22.0**	**18.2**	**34.7**
Oil				
Domestic Petroleum Supply	**9.7**	**7.8**	**26.4**	**12.3**
Imports				
— North America	**0.3**	**4.0**	**0.8**	**6.3**
— Other	**7.8**	**2.0**	**21.2**	**3.1**
Oil Shale	**0**	**2.4**	**0**	**3.8**
	17.8	**16.2**	**48.4**	**25.5**
Natural Gas				
Domestic Supply	**9.0**	**6.9**	**24.4**	**10.9**
Imports				
— North America	**0.4**	**1.2**	**1.1**	**1.9**
— Other	**0**	**1.0**	**0**	**1.6**
	9.4	**9.1**	**25.5**	**14.4**
Nuclear Power**	**1.4**	**13.3**	**3.8**	**21.0**
Hydroelectric	**1.5**	**2.1**	**4.1**	**3.3**
Geothermal, Wood & Waste	**0**	**0.2**	**0**	**0.3**
Solar	**0**	**0.5**	**0**	**0.8**
	36.8	**63.4**	**100.0**	**100.0**
Domestic			**76.9**	**87.1**
North America			**78.8**	**95.3**
Overseas			**21.2**	**4.7**

*1 million barrels of oil equivalent/day (MMBOE/day) at 5.8 million Btu/barrel = 2.117 $\times 10^{15}$ Btu/year or 2.117 quads/year.

**Assumed 450 GWe installed capacity in the year 2000.

that predicted by most other investigators. We feel that this is likely because production of coal for domestic use will be largely demand limited until at least the late 1980s, the earliest date when synthetic fuel production from coal and expansion of coal-fired electric power capacity appears likely to have a significant impact on coal demand. However, total U.S. coal production is expected to grow more rapidly than domestic consumption in order to satisfy the demand for export steam coal, which could provide a partial substitute for oil in international commerce. The foreign exchange revenues from such exports would contribute toward offsetting the dollar outflow to finance oil imports.

The 450 GWe of nuclear power is appreciably above the current Department of Energy estimates, yet well below the industry's estimated potential for bringing 550 GWe on-line by the year 2000.* Considering the negative impact of the Three Mile Island incident, however, bringing 450 GWe on-line by 2000 might be difficult to achieve. On the other hand, the substantially higher risk of continuing the current heavy dependence on oil and gas is gradually being recognized, and we remain convinced that nuclear power will play a major role.

Energy Prices in the Year 2000

To estimate the average delivered price of energy to consumers, the energy sources shown in Table 2 were regrouped to determine the distribution of energy use forms such as electricity, liquid fuels, and gaseous fuels (see Appendix A). The specific sources of synthetic fuels were identified as well; assumptions were then made as to the relative production of refined products from liquid fuels.

The real cost of electric power was assumed to escalate at 1 percent per year, comparable to the rate of 1.1 percent per year estimated in a recent Arthur D. Little Study.** Coal prices and conventional petroleum prices were assumed to escalate at 1 percent and 2 percent respectively over the basic rate of inflation. Prices for synthetic fuels from coal and oil shale were obtained from recent estimates presented to the National Academy of Engineering.***

Rough estimates of the prices of various end-use energy forms were combined with their expected relative consumption to yield average energy prices in 1978 and 2000. For the particular fuel prices and fuel mix assumed, average real price of energy *delivered to the consumer* in the year 2000 was estimated to be $36.1/barrel of oil equivalent ($6.22 per million Btu) or 51 percent above 1978 levels. Since GNP in constant dollars is assumed to grow by 92 percent (3 percent per year) during this period, however, the total cost of energy per unit of GNP would increase

---

**Nuclear Power: Its Potential and Resource Needs,* Atomic Industrial Forum, Inc., published April 1979.

***Implications of Low Electric Power Growth through 1990,* Arthur D. Little, Inc., February 1979.

***At the Academy's Annual Technical Meeting on Hydrocarbon Fuels, November 2, 1978. Reported in *The Bridge,* Vol. 8, No. 4, Winter 1978-79, pp. 12-20.

from 12.2 percent in 1978 to 14.5 percent in 2000, about a 20 percent increase.

Conclusion

By restructuring the pattern of energy supply toward increased use of its abundant domestic resources—coal and uranium—the U.S. could regain substantial control over the security and cost of its energy supplies by the year 2000. If it should become necessary, nearly complete energy independence might ultimately be achieved, possibly as soon as 2010 to 2020. Yet, time is running out and action is urgently needed to formulate and implement a realistic plan to reverse the steadily deteriorating energy situation. Accomplishing this goal is virtually impossible under present conditions where existing and proposed legislation and regulations appear largely designed to decrease the supply of energy and to increase costs to a prohibitive level.

Based on the hypothetical mix of energy sources used in this analysis, we believe that if the nation elects to take positive action, the resources and the technology will be available to meet future U.S. energy needs with appropriate environmental safeguards at a cost only about 50 percent above the cost of energy today. Although the actual proportion of the various energy sources may vary substantially from the mix assumed, the conclusions would not be appreciably different, if both coal and nuclear power are available to provide a major portion of future energy needs. We hope that an objective assessment of the risks of the present course of energy policy will lead to a positive approach along these lines to ensure the future well-being of the nation.

In the preparation of this paper, Mr. Davis acknowledges with thanks the assistance of R. O. Sandberg.

Appendix A
Energy Use Forms

	MMBOE/Day* 1978	MMBOE/Day* 2000	% of Total 1978	% of Total 2000
Electric Power				
Coal-Fired	4.9	14.1	13.3	22.1
Nuclear	1.4	13.3	3.8	21.1
Hydroelectric	1.5	2.1	4.1	3.3
Oil and Natural Gas	3.1	0.9	8.4	1.4
Geothermal, Wood & Waste	0	0.2	0	0.3
Solar-Electric	0	0	0	0
	10.9	30.6	29.6	48.2
Coal *(Excluding Electricity, Synfuel Production)*	1.8	3.7	4.9	5.8
Liquid Fuels				
Petroleum				
Domestic	7.9	7.3	21.5	11.5
Imports				
— North American	0.3	3.0	0.8	4.7
— Other	7.8	2.0	21.2	3.1
Synthetic Liquids				
Domestic *(Coal-Based)*	0	1.8	0	2.8
Canadian Imports	0	1.0	0	1.6
Shale Oil	0	2.4	0	3.8
	16.0	17.5	43.5	27.6
Gaseous Fuels				
Natural Gas				
Domestic	7.7	6.5	20.9	10.3
Imports				
— North American	0.4	1.2	1.1	1.9
— Other	0	1.0	0	1.6
Syngas from Coal	0	2.4	0	3.8
	8.1	11.1	22.0	17.6
Solar Energy *(Non-Electric)*	0	0.5	0	0.8
	36.8	63.4	100.0	100.0

**1 million barrels of oil equivalent per day.*

Appendix B
Electricity Production and Growth

We have found that electricity production over the past 25 years or so can be correlated as a function of GNP, electricity price, and time by the equation:

$$E = \frac{7.114 \times GNP \times e^{-0.1177p}}{1 + 6.659e^{-0.03335T}}$$

E is the annual production of electric power (kwhrs/year) by electric utilities (excluding industrial generation). GNP is the gross national product and p is the average delivered price of electricity to all utility customers in cents per kwhr, both expessed in constant 1972 dollars. T is relative time in years, with T = 0 in 1947. A comparison of calculated versus actual electricity production is shown in Figure 2. While this relationship is largely empirical and its forecasting

Fig. 2 Historical Relationship Between Electricity Production, GNP, and Electricity Prices

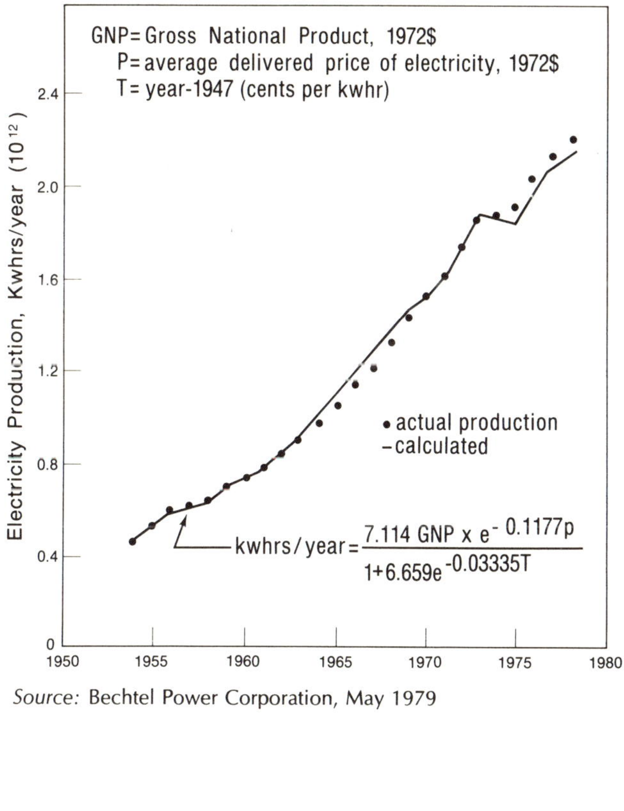

Source: Bechtel Power Corporation, May 1979

properties are open to question, it was used to calculate average electricity growth rates between 1977 and 2000 for various of increase in GNP and real electricity prices. The results are shown in Table B-1.

Table B-1. Electricity Growth Rates (1979-2000), %/year

Increase in Real Electricity Prices 1979 - 2000 (%)	Growth Rate of Real GNP (%/ year)		
	2.5	3.0	3.5
0	4.6	5.1	5.7
25	4.3	4.8	5.4
50	3.9	4.4	5.0

Discussion

Chairman: One approach that has been suggested to spur production of synthetic fuels is for the government to announce that it will contract to purchase perhaps two or three million barrels a day of such fuels at the lowest price at which it can get a bid, presumably something on the order of 150 percent of today's prices, thus involving private capital at risk, seeking a return. Is that plan, in your opinion, at all practical, and if not, what stands in the way?

Davis: It is a possibility that certainly should be considered, and might be useful in some cases. The first thing that occurs to me, having worked for the government and having negotiated with them, is that I am not sure how a contract like this would be written. Perhaps the first step would be on a scale to get some of the technology started—which might be substantially less than two or three million barrels a day—specific projects where the costs and the basis for the price could be identified. Other ways have been suggested, of course—various kinds of tax incentives that would be somewhat more automatic and still leave a basis for more competition among various companies. Another idea that has always intrigued me is a requirement similar to that we have had on mileage requirements for automobiles, that is, the government would require that a certain percentage of each gallon of motor fuel be derived from synthetic sources, and simply let the supply companies figure out how they can best meet this schedule. That is perhaps a little complex, but it may be less complex than some of the other schemes.

All these things have a basic problem, however, that I think would bother industry (or parts of it) to a substantial extent, and that is, how is it assured that the government is going to live up to the promises and contracts it made? With the vacillations and changes in the political scene and the Congress, I am not sure bankers would be eager to put up large sums on an agreement by the government to pay a lot of money ten, twenty, thirty years downstream.

From the floor: You have described a scenario of rapidly increasing capital intensity of the energy industry. And you have talked about synthetics—liquid synthetics from coal—something not quite on the order of magnitude greater than the present petroleum system, but close. Do you have any views on how the necessary capital formation might take place, or on the adequacy of the U.S. manufacturing industry that would have to grow and supply two million barrels per day by 1995? You are talking about perhaps 200 to 250 billion dollars to achieve this two million barrels per day of synthetic fuels from coal. Do you have some view on the adequacy of our supply system?

Davis: The plant investment, I think, turns out to be a major part of it, at least our studies show. The present capacity of the engineering construction industry has put in place some 75 to 100 billion dollars of new construction per year; and for nuclear power plants, the figure is something on the order of 30 billion dollars per year.

So these synthetic fuels numbers, compared to some other types of investments, do appear large; it may be necessary to use priorities. But I do not think what is being discussed is beyond the physical capacity of the manufacturing and engineering construction industries, providing it is not tried to do everything at once by giving a high crash priority to everything.

The problem is more one of how to get the industry in financial shape so bankers will provide the money. Again, the financing, presumably, is there. In fact, my friends from the financial community say it is all a matter of price and priorities. But today, and particularly if one looks at the utility industry, for example, nuclear power is considered a very bad risk by the financial people. So is coal. The reason largely is uncertainty, and I stress this word, *uncertainty,* on licensing, on operation, on cost, on revenues—everything associated with this is uncertain today. And if it were my money, I am not sure I would invest it that way either. What is needed is a system where there is continuity and some certainty to it.

From the floor: If a synthetic fuels program were to be initiated by the mechanisms that have been discussed and some of the other uncertainties could be resolved, what about *future* uncertainties with respect to environmental requirements?

Davis: In my view this is one of the major areas of uncertainty affecting the electric utility industry or making fuel from shale oil; it is a major consideration that has not been faced yet in making synthetic fuels from coal. I see no real way to deal with it other than to put certainty into the process somehow. At least there should be the conclusion that once a plant has been built and is in operation, the ground rules for that plant are not going to be changed from time to time, unless it can be demonstrated that there is something fundamentally different that was not considered and is sufficiently important to require major changes. This business of having to retrofit and change everything after the fact is simply almost impossible for owners to consider and financial backers to take into account. There are certain risks that people are willing to take; some others are just beyond the pale.

From the floor: I wonder if perhaps the industry is not in effect abandoning the free market supply/demand system and relying excessively on government to provide support for industrialization of new technologies.

Davis: Several things are important in industry-government relations, and one major area of controversy is confusion in the licensing and regulatory requirements, and since they are imposed by the government, that basically can be resolved only by the government, whether it be at the federal, state or local level.

As for allowing the economics of supply and demand to lead to initiation of production in accordance with the requirements, I think that it is, first of all, limited by uncertainties with respect to licensing and regulation and revenues that we now have; but second, I think the problem that bothers many of us in industry and is involved with what government can do either directly or indirectly is simply the enormous length of time it takes to develop and deploy a new large-scale energy technology.

We are talking in terms of perhaps fifteen or twenty years needed to bring about a synthetic or shale oil capacity of two million barrels a day, even with the technology in hand today. It would require an absolutely enormous program and a narrow area to do it in a very much shorter time. These time periods are longer than the time periods in which the economic situation is changing, so given a situation without long-term guarantees and assurances (which I think are almost impossible to have) and given an estimated need for two, three, four, five million barrels of oil a day from gas from synthetic sources by the year 2000, no-one is going to rush out and start building those plants. The economics are not attractive enough now, even though we may be sure they are going to be attractive in 2000.

So the question is, how to get this started? This is, in my opinion, a fairly legitimate function of government—to get something started that is going to be desirable in the long run when there is no basis for getting it started today, and where eventually the supply and demand and price considerations might lead to it, but possibly far too late to be useful. I think we are headed for a situation where there will not be time to solve the problem. That is my perception of it.

From the floor: Solar energy cannot make a significant contribution between now and the year 2000, yet does not the answer to your last question imply that no new source like oil shale or geothermal could make a significant contribution in that time frame either, because of the kind of time restraints you just identified?

Davis: Not necessarily. Today we do not know even how to produce solar energy on an economic basis that would allow it to make any contribution. If we had the technology in hand, it probably could be done by the year 2000; but it does not exist. In my opinion, we would be hard pressed within the next twenty years to develop several million barrels a day equivalent from *any* alternative source, even assuming we have the know-how today. With solar, we do not even have the technology yet.

Dale D. Myers

Expensive, Yes–Plentiful, Maybe

The theme of this conference is the premise that America can have prosperity with plentiful but higher priced energy. "Higher priced" is clear. "Plentiful" is less definite. The United States has plentiful domestic resources, but many technological, economic, and environmental uncertainties stand between those resources and their use for energy.

- Nuclear — safety, proliferation, and nuclear waste
- Coal — strip mining, air pollution, and possible "greenhouse" effects from CO_2
- Oil shale — expensive and with air and water pollution
- Solar — expensive, intermittent, and with major technological barriers to overcome.

But as the Department of Energy looks out into the uncertainties of the future, one thing is clear: the United States can do much more to assure that its plentiful energy resources are brought to market. These supplies will cost more than the current, subsidized supplies. However, the alternative to these higher prices would go beyond the simple price of energy: reduced economic growth, unemployment and a loss in na-

---

Under Secretary of the U.S. Department of Energy from 1977 to 1979, **Dale D. Myers** *is an engineer with an illustrious career in astronautics. As associate administrator of manned space flight for NASA, he twice earned its Distinguished Service Medal. His earlier work with North American Aviation/Rockwell International included missile development and the space shuttle program, and led to a Corporate Vice Presidency. Myers is a national director of the American Institute of Aeronautics and Astronautics, and a member of the National Academy of Engineering.*

tional security. It will be useful to consider some of the things that can be done to avoid these costs.

Historical Perspective — Where Have We Been?

It is interesting to note that almost 130 years ago a patent was issued in Scotland for a process to convert coal into oil. When the patent was issued, the world was running short of supplies of whale oil and candle wax which had been the main sources of light. As a result, coal oil appeared to have a promising future as a new source of energy.

In 1854, four years after that patent was issued, George Bissell, a young lawyer from New York, became interested in the possibilities of using crude oil as a source of kerosene. In fact, he was so impressed that he formed a company and purchased some land in Pennsylvania where crude oil seeped through the ground and could be collected in modest quantities. The local coal merchants paid him $20 a barrel (a price not repeated for 100 years) for his product. In 1859, he hired a man named Drake to manage his rather badly run company, and on August 27 of that year "Drake's Folly" struck oil; the world's first oil well began producing at the rate of thirty barrels a day.

The Drake discovery began to feed the nation's apparently insatiable appetite for oil and other energy sources. The United States was about to begin a century of rapid economic growth, fueled in part by its abundant energy resources; coal superseded wood, and oil and gas superseded coal. However, increased energy use and increased economic growth did not move in lockstep fashion during this period. Energy use rose much faster than GNP during the last part of the 19th century and through the first part of the 20th century; but since that time the nation has been able to increase its output of goods and services without using as much energy per unit of GNP (Figure 1).

Between 1950 and the present time, there have been some dramatic changes in energy consumption. With the exception of the period between 1965 and 1970, energy consumption per unit of GNP has declined steadily. In fact, after adjustment for the 1965 to 1970 period, energy consumption has grown at only half the rate of increase in GNP (Figure 2). The conclusion to be drawn is that an improved standard of living need not be tightly tied to buying and using more energy-consuming goods.

This conclusion is important as we develop energy policies for the nation because it means energy supplies can be used more efficiently while development begins of our abundant domestic resources—coal

Fig. 1. U.S. Energy Consumption per Unit of GNP: A Historical Perspective

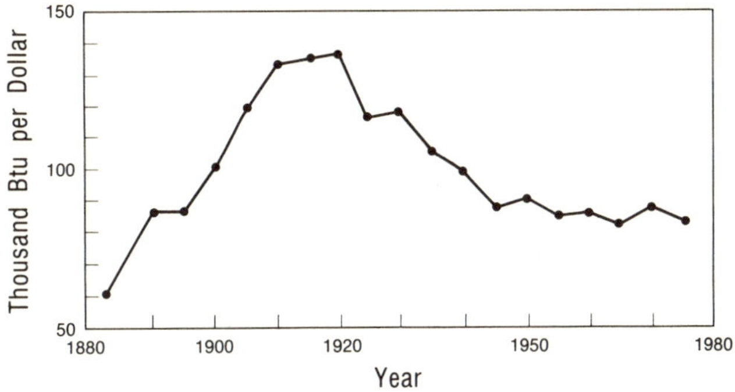

Source: Sam. H. Schurr, "Energy, Economic Growth, and Human Welfare," ERPI Journal 5(4):14-18 (1978). Reprinted with permission.

Fig. 2. U.S. Energy Consumption per Unit of GNP: Post-World War II

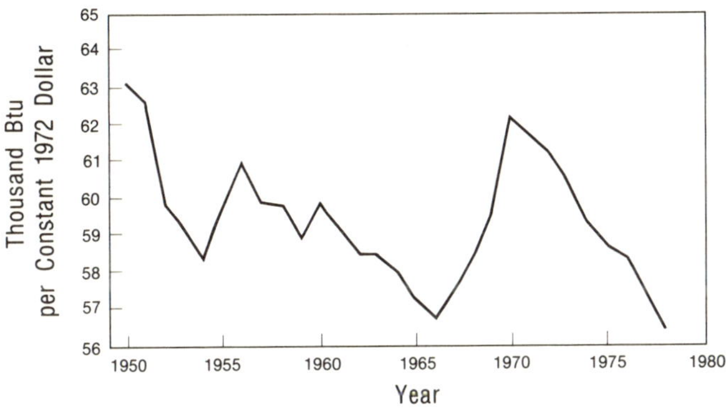

and nuclear power as well as renewable and virtually inexhaustible sources of supply—solar energy and fusion.

The Energy Future

While it is reassuring to know that GNP growth is not dependent upon comparable growth in energy supplies, that knowledge does not directly provide an answer to the U.S. energy problem: *the United States is too dependent upon oil.* Over the near term this problem manifests itself in situations such as the recent Iranian disruption. As the currently tight supplies of gasoline demonstrate, the United States is highly vulnerable to supply interruptions.

The Iranian crisis, however, is only a symptom of a broader longer-term energy problem facing this nation and the world. The world's oil supplies are limited and they cannot be increased fast enough to keep pace with rising demand. Although the time when worldwide oil production will peak cannot be predicted precisely, it is unlikely that even known quantities of oil can be developed in time to offset the projected declines in production that will occur during the rest of this century.

Conventional oil and gas production worldwide is expected to continue to decline in the 21st century. Unconventional supplies and transitional supplies of liquids and gases will have to supplement conventional production, and it is likely that eventually, renewable and virtually inexhaustible sources of supply will become major factors in the nation's energy balance.

The challenge for energy policy-makers is to ensure an appropriate balance between improved long-run security of supply and an ability to deal with sudden crises, such as Iran.

These problems have been addressed during the preparation of the President's second National Energy Plan. I would like to share with you some of the analyses and conclusions contained in the Plan.

The Near Term

Over the next seven years, 1979 to 1985, the United States and the rest of the world will be fortunate to escape a second large increase in world oil prices. If prices should rise significantly, the adjustment process would again be painful, for the United States could do little to influence oil supply and demand.

If world oil prices are held constant, U.S. primary energy demand should increase at about 2.2 percent annually through 1985. Such a rate would be higher than the 2 percent sought in the first National Energy Plan but much lower than the 2.8 percent that would have occurred without the 1978 National Energy Act.

Energy conservation measures taken in response to the quadrupling of prices that have occurred since 1973 and the tax credits contained in the National Energy Act account for the relatively low rate of energy growth expected over the next seven years. Despite this improvement in the efficiency with which we use energy, Figure 3 shows that the "import gap" between U.S. energy consumption and domestic production will

Fig. 3. U.S. Energy Consumption in the Near Term
(Medium World Oil Price Case)

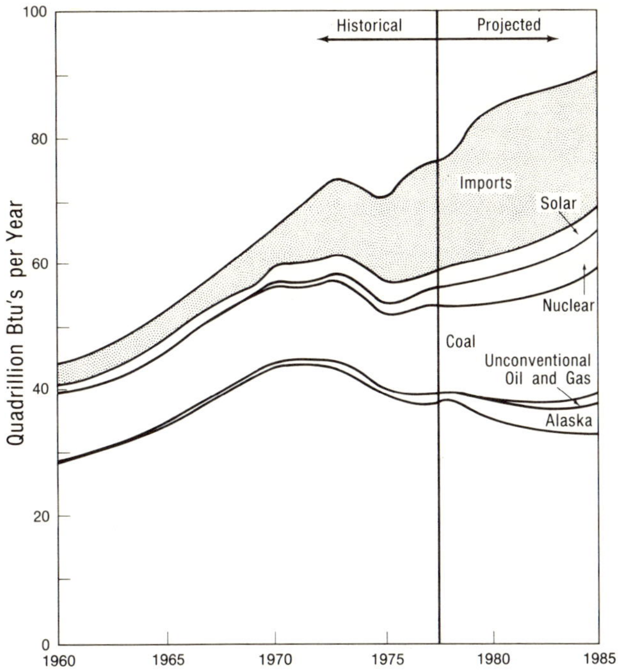

With the President's program, imports remain near current levels in the near term.

continue in the near term at nearly a constant level! If imports measure a nation's vulnerability to supply disruptions and price increases, then U.S. vulnerability will remain high during this period and the country can anticipate continued insecurity from these sources.

There are, however, some things that can be done to reduce U.S. dependence on foreign oil and vulnerability to supply disruptions.

- *Energy must be priced at its true "replacement cost."* With the phase-out of artificial price controls, market forces should lead naturally to more investments in conservation and production. The President's oil pricing decisions concerning newly discovered oil, enhanced oil recovery, and production from marginal wells, as well as the phased removal of controls on domestic crude oil, could result in an increase of 750,000 barrels per day above the levels that would have occurred with continued controls. These new supplies, plus the conservation measures induced by the increased prices for domestic crude oil, should enable the country to hold its imports at current levels through 1985. Please note that all of the above only hold imports constant.

- A second action that can be taken over the near term is the *elimination of excessive regulatory delays and needless institutional barriers* that paralyze construction of refineries, pipelines and other energy facilities. Construction of the Sohio Pipeline in California (which required 700 permits), the oil refineries in Maine and Virginia, and power plants such as the Sun Desert Project are examples of situations where conflicts between national, state and local priorities need prompt resolution. Similar uncertainty exists about current and prospective air quality standards which have dimmed the attractiveness of coal. Recently enacted legislation requires all new power plants to use coal or other alternate resources, and government now must provide a stable set of environmental rules and work with industry to develop environmental control technologies to meet the new standards. Stable standards are the key, and they have not been provided yet.
- The third part of U.S. strategy would ensure that government regulations provide the *proper incentives for increased production* of crude oil and refined products. Over the near term additional crude oil production can be encouraged through the use of enhanced oil recovery techniques to increase yields from known reserves. Regulatory incentives should also encourage investments in upgraded refinery capacity to prevent shortages of unleaded gasoline and other products most in demand.
- Finally, because near-term domestic production cannot displace imports altogether, the United States should seek to *diversify its sources of import dependence and enhance its security of supply.* The United States should support multi-lateral assistance to increase production in non-OPEC countries and encourage immediate efforts to determine the potential oil resources that various non-OPEC countries possess.

The United States can also enhance the security of its supply with the strategic petroleum reserve. A billion barrel reserve with the capability of replacing all the oil denied the United States during most embargoes and short-term interruptions would allow the nation time to resolve the problem causing the embargo while postponing the economic distress that would arise immediately from a large oil shortage. By delaying the economic effects of any international disruption, the strategic petroleum reserve can serve as a powerful strategic deterrent to foreign adversaries.

These measures, in addition to the tax, pricing, and regulatory measures included in the recently passed National Energy Act, constitute one near-term strategy and also set the stage for policies that buy even greater energy security in the midterm.

Midterm

At some point before the end of this century, it is likely that world oil production will peak and begin to decline. The result will be uncertainties about price and availability of supply. In part because of these uncertainties, the nation must begin to reduce its reliance on conventional oil and gas, and undergo a transition as it begins to develop alternative sources of supply. If it manages this transition well, it can maintain a healthy rate of economic growth; if it does not, the nation will pay for the higher cost of energy in reduced employment and output of goods and services.

It is not possible to predict precisely when the peak in oil production will occur. That will obviously depend on the rate of discovery and development of new oil sources and the rate at which existing production declines. Oil from existing world fields is expected to start declining around 1990, so as much new oil will have to be discovered and produced by 2000 as was produced in 1970 (Figure 4) to meet even a 2.5 percent average annual growth rate in world oil consumption. Even if large new fields can be discovered and developed by the early 1900s, the incremental production from such fields would be swallowed up quickly by demand growth.

In planning for the midterm, the Department of Energy has tried to put some bounds on the uncertainties that characterize this period by using three paths for world oil prices through the end of the century (Figure 5).

The range of price uncertainty is due not only to the uncertainty over when oil production will peak, but also to uncertainties over other sources of supply. Under the three world oil price cases, oil and gas production varies little from current levels; coal is counted on at 2.5 times 1977 levels and nuclear power at 5 times 1977 levels to provide most of the increase in supplies through the rest of this century (Table 1). If technological, environmental, or institutional factors constrain pro-

Fig. 4. Oil Required to Maintain a 2.5 Percent Annual Growth in World Oil Consumption[a]

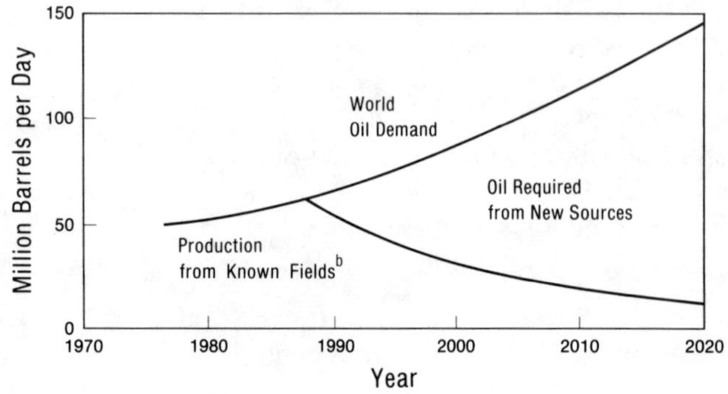

[a]Does not include centrally planned economies.
[b]Assumes 600 billion barrels of known reserves and a 5% per year decline in production from developed fields.

Fig. 5. Three Paths for World Oil Prices

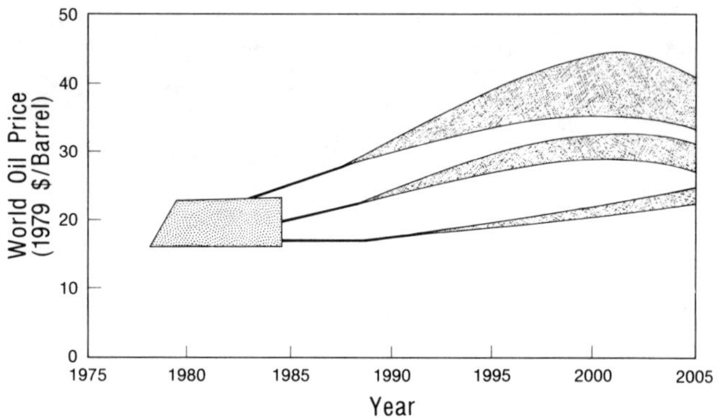

duction from these sources, there would be increased demand for imports with resultant pressures on world prices.

During this period, the rate of growth in energy consumption is likely to slow to an average annual rate of about 1.8 percent per year. Surprisingly, energy consumption is much more sensitive to economic growth than to oil prices. Higher oil prices discourage overall consumption but tend to encourage more use of electricity and other energy-intensive fuels. Lower prices encourage increased consumption and the direct use of oil supplied by imports.

Table 1: Primary Energy Supply in the Year 2000
Effect of World Oil Prices *(Quads per Year)*[a]

	1977	*Year 2000 Projections* Low Prices	Medium Prices	High Prices
Domestic Production				
Crude oil and NGL	**20**	**20**	**21**	**22**
Natural gas	**19**	**18**	**18**	**19**
Coal[b]	**14**	**33**	**36**	**40**
Nuclear	**3**	**15**	**16**	**17**
Hydro, Solar, & Geothermal[c]	**4**	**9**	**10**	**10**
Total Production	**60**	**95**	**100**	**106**
Imports				
Oil	**18**	**27**	**18**	**10**
Gas	**1**	**2**	**2**	**2**
Total Imports	**19**	**29**	**20**	**12**
Total Consumption[d]	**78**	**123**	**119**	**117**

a. Total may not add due to rounding.

b. Net of approximately 2 quads of coal produced domestically and exported.

c. Includes 1.8 quads of decentralized biomass currently unaccounted for in Department of Energy statistics. These balances account for decentralized solar as *end-use* quads displaced. For example, in the medium price case, 3 quads of decentralized solar penetration represent 4.3 quads of *primary* energy displaced. The medium price case, therefore, represents 11 quads of primary solar energy in 2000.

d. Net of approximately 1 quad of refinery and other losses.

Although the energy consumption impacts of differing price cases are relatively small, the economic impacts on the nation could be quite large under the high-price case. Compared to the low-price case, high prices would reduce GNP in the year 1995 by $100-200 billion, 2.5 percent to 5 percent of expected GNP. Inflation would increase by .3 percent to .6 percent. In sum, the nation would be significantly less well off than were oil price increases to follow the low or moderate price path.

The nation can minimize its exposure to the high price path by building on the replacement cost pricing policies initiated over the near term. Replacement cost pricing will encourage maximum conservation and lead U.S. consumers to meet more of their demands from sources other than oil imports. Artificially low near-term prices would not only quicken world demand but also slow the development of alternative supplies. Eventually, the underlying pressures would prevail and prices would have to rise abruptly to balance demand with supply.

Coal and nuclear power are critically important transitional fuels in the midterm. A broad strategy must be developed to provide a consensus on the best approach to the environmental, safety, logistical, and regulatory problems associated with these fuels. Slippage in production of either of these sources could intensify the pressure on oil markets and further increase prices.

Third, the United States must turn increasingly to unconventional oil substitutes to reduce the risks of high oil prices and their resultant economic loss. If these unconventional energy supplies—direct petroleum substitutes, such as heavy oil, tar sands and synthetic liquids and gases as well as indirect substitutes, such as some solar technologies—were available in sufficient quantities, they could help hold down world oil prices by providing a backstop and by easing oil demand pressures. Although the impact of these backstop technologies on total production would be small during the midterm because of the long lead-times to bring them into production, they could exercise some constraint on world oil prices by reducing U.S. oil imports (Figure 6).

The transition to new unconventional technologies and high cost energy sources will take many years. These technologies will require enormous investments over a long period of time. To pursue them too quickly could be more costly for the economy than the problem they are intended to address. If world oil prices follow the high price path and reach $30 a barrel by 1990, the United States is already behind in developing new oil substitutes. If world oil prices follow the low price path, the United States has been prematurely investing in such technologies.

Current strategy as reflected in the Department of Energy's budget coincides with the level appropriate for a medium price path. By the late 1980s, for example, the first commercial plants for synthetic liquids would be on line with product costs close to the prevailing world oil price. No-one can be certain how fast or how slowly world oil prices will rise. Prudent strategy, however, will plan for the pessimistic cases which, even if unlikely to occur, would be extremely damaging to the economy. With the serious national security implications of even short-term supply interruptions, I believe heavy government investment in liquids from coal and oil shale is critically needed.

The Long Term

The United States faces two major transitions in energy markets between now and the middle of the 21st century. The first will occur in

Fig. 6. Production from Backstop Technologies in the Midterm
(Medium World Oil Price Case)

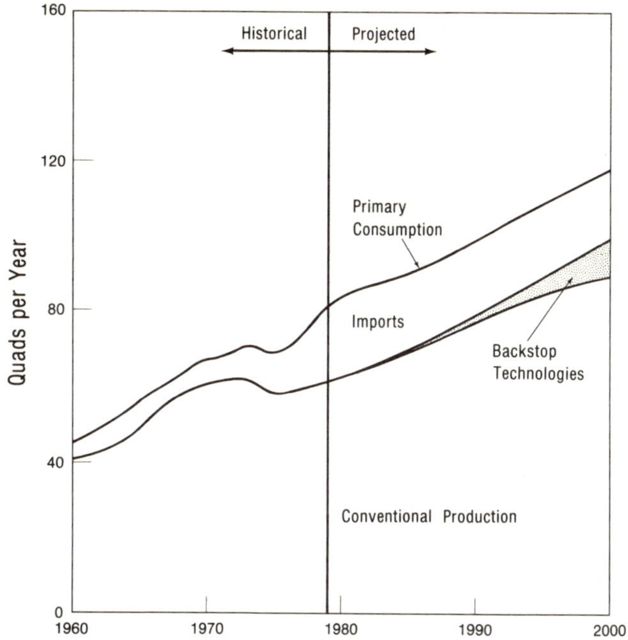

U.S. Backstop Technologies include:

- Shale Oil
- Synthetic Liquids
- Unconventional Gases
- New Solar Technologies

Production of backstop technologies will make only a small contribution in the midterm.

the midterm period when the country moves from an energy system tied to traditional oil and gas sources (including imports) to one relying on unconventional sources. The "transitional" energy supplies include a small contribution from a few cost-effective renewable technologies, but mainly consist of finite, non-renewable energy sources that are not yet in wide use. Examples include enhanced oil recovery, oil shale, coal-derived liquids and gases, unconventional gas, and conventional nuclear energy. If successfully deployed, these higher-cost sources will reduce the nation's import dependence without sharp economic disruptions or mismatches between the new energy forms and end-use requirements.

But however large they may seem, even these transitional energy supplies are depletable. For the mid-range projection, recoverable resources of coal and uranium appear sufficient to sustain the expected increases in electricity and other delivered fuels only until the first half of the next century, at most. Because of the depletable nature of these transitional fuels, *as its long-term objective the nation must bring on renewable and essentially inexhaustible sources of energy to sustain a healthy economy.*

A second transition, therefore, would start after the year 2000. A set of "ultimate" technologies, including all the renewable and advanced nuclear technologies, would begin to displace traditional fuels and non-renewable conventional supplies. These new sources include the various solar technologies, hydroelectric power, and the advanced nuclear technologies, such as fusion and breeder reactors, which are essentially inexhaustible supplies. I also want at least to mention hydrogen as one of the possible long-term sources of energy.

Whether the United States reaches the 21st century with the full conservation ethic or not, it will be moving toward these ultimate sources of energy. Traditional domestic energy sources (including onshore, offshore, and Alaskan oil and gas) peaked in about 1970. Non-renewable transitional sources are likely to peak sometime around 2020. Ultimate energy sources probably will not dominate the energy picture until around 2070. Depending upon future energy growth and costs of ultimate energy sources, the transition to ultimate sources should be underway by 2020. Ultimate sources could grow from near negligible amounts in 1975 to about 10 percent of domestic primary energy consumption by 2000.

The uncertainties about resources and prices in the 21st century dwarf any that have been discussed previously. It seems likely, however, that a first group of transitional technologies will slow or stop the oil price increases during the 2000-2020 period. If unconventional supplies are held back by environmental or institutional problems, a new and even larger explosion in world oil prices would occur, going well beyond the three price cases projected. Indeed, world oil prices would continue to rise until halted by substantial use of renewable and advanced nuclear technologies later in the 21st century.

The Department of Energy currently supports through its R&D efforts development of three major long-term energy sources—solar energy, nuclear fusion, and breeder reactors. It is uncertain which technologies will constitute the base for post-2000 energy supplies. The costs, the environmental impacts, and the technical processes themselves are not well understood in many cases. For that reason, if no other, federal research and development should not focus narrowly on one technology to the exclusion of others.

Conclusion

In conclusion, I go back to my original premise: the United States has ample energy resources. It needs only the will to develop them. The production of these resources as energy supplies will cost more than prices now being paid by the American consumer, but the cost will be cheap in comparison to the alternative of not developing them. So to the premise of this conference, I say, expensive, yes; plentiful, only with deliberate and farsighted action, now!

Discussion

Chairman: Mr. Secretary, if the Department of Energy were called upon by the President of the United States in a Pearl Harbor sort of crisis to set up a synthetic fuels industry, would you have confidence that it could in five years have synthetic fuel flowing at a rate greatly larger than what you have been talking about?

Myers: There is no question about that. As things stand today, this whole business of developing energy is one of the rights of the private sector. The Energy Department has development programs to alleviate the risk level; we will build a single plant as a capability rather than a capacity, to show for the Department and for the industry what the cost is, what the products are, what the reliability in on-line capability of the plant is; but we do not really build capacity.

If DOE went into a capacity building operation and production business, instead of having budgets of three billion dollars for R&D, we would be building billion dollar plants, first at the rate of one or two a year and then at the rate of ten or fifteen a year. And we would be moving at a rate of production like that of the private sector with nuclear power plants. Presently, we are building plants at the rate of ten to fifteen a year in that area, too.

Yes, the Department of Energy could bring synthetic fuels on in a hurry. But it is interesting to note that, even under those circumstances, in a period as short as five years there probably would not be sufficient capacity out of liquids-from-coal even to offset the oil that is being imported. Ten of these liquid plants would equal one quad, and one quad is 500,000 barrels a day, so it would take quite a while actually to offset U.S. imports.

From the floor: Can we manage the capital-intensive transition without a very sharp drop in personal consumption standards?

Myers: I think that it would not make a significant difference. Most of these plants, if the price of oil goes up to $25 a barrel, are going to be payback situations. The only reason private industry is not moving into these plants now is that they have great uncertainty, as does DOE, as to the price of oil. By getting one or two of these plants on line, as the Department is doing now, we are prepared for '83, '84, when the price of oil will be high enough that industry will be ready to invest. I think that it will have very little effect on our standards.

From the floor: After the Energy Act of 1978, there was a 15 percent drop in drilling rig utilization. Is there any chance to modify this onerous act and deregulate natural gas, which is in relatively abundant supply?

Myers: Natural gas is now being deregulated, and I think there has been an impact. There will be a little trouble with drilling rigs because we are also deregulating oil and most of the rigs in the country are busy right now. Until enough rigs are produced, it is not possible to catch up with exploiting all the capability in this country.

From the floor: There are some, including some here, who think that oceanthermal and geothermal can supply a much larger fraction of the requirements than you have indicated. Would you like to say something more about those?

Myers: Geothermal has been a really good resource in California, in the geysers area, with steam. There are plants now being built down in the Imperial Valley and in New Mexico; there is some geothermal up in Utah; but there is not much in other places in the country. In total capacity it is a very limited resource. Still it looks like an attractive resource and it is really being exploited, a little by the Department of Energy, and quite a lot by the private sector. We are delighted to see it moving out of the Department into the private sector.

OTEC is another matter. This is a very high technology area working on a very small *delta T,* a small thermal difference between deep oceans, cold waters, and warmer waters at the surface, and some of those old laws of thermodynamics still almost apply. The Department will probably do an R&D program just to see exactly where we stand in it, but I really do not see OTEC as a major contributor unless there is a lot of magic in that system that is not evident.

Chairman: I will put together a number of questions that have come from the audience: Gasohol . . . fuel from grain and other agricultural products . . . mandating that a certain fraction of the fuel sold by companies should consist of synthetics—that whole area. Would you care to comment on those?

Myers: Gasohol is the kind of grain alcohol that is used for vodka, and it costs about $84 a barrel. Using waste agricultural products as the heat source to distill other waste agricultural products has brought alcohol down into the $40 or $50 range. You may not know that the government is actually subsidizing gasohol. We took off the 4¢ excise tax on gasoline that is 10 percent alcohol, and that amounts to a $16 per barrel subsidy, so it is beginning to go. That excise tax has been taken off until 1984, but that is not long enough for somebody to build an alcohol plant and get his money back on it. For that reason, the new energy plans suggest its removal until 1994. You will see a lot of waste agricultural stuff turned into grain alcohol. A better way to go would probably be methenol, which is cheaper; someday that will come into the system.

The Department of Energy is looking at the idea of requiring a certain percentage of gasoline to be synthetic. There are a lot of institutional problems in developing such a system but it is being looked at very seriously, and DOE is trying to find some way to press the industry into moving into the synthetics rather than having the government be the producer.

From the floor: The forecast that you gave us obviously hinged on greatly expanded dependence on nuclear power. In view of the extreme negative reaction to nuclear as a result of the Three Mile Island accident, do you think that is at all realistic? Does the government believe that it is possible to bring into implementation the kind of expansion of nuclear energy that is in the National Energy Plan?

Myers: I talked about a 5-times increase in nuclear power. The United States has a little more than 50 gigawatts of nuclear power on line today. And the number that Kenneth Davis has mentioned is something like 450 by the year 2000; that would be a 9-times increase. The Department projects about 5½-times increase.

The Department of Energy sees no alternative to nuclear power and coal. Admitted, there are difficulties with nuclear in terms of safety and personnel training and many other things that are going to be given excruciatingly detailed attention in the system—I have already seen it in the utilities. And I think it will all be to the good. Many times in our national history the United States has gone through crises and, after coming out of them, has had a better situation. And I think, after going through this business of understanding the issues that were involved in that accident and correcting them (and you can be sure they will be corrected)—making the detailed design changes to the facilities, detailed changes to the control room operations, to the training of the operators, and all the rest—the country will come out ahead as far as safety in nuclear power is concerned. We are in a position where we really have no choice [about staying with nuclear]; and I think we are going to have a safe power source in nuclear energy that we will all be very, very happy we have.

From the floor: To what extent can conservation programs such as mass transportation, shifting from truck to railroad transportation, etc. offset the growth of domestic oil consumption trends?

Myers: We have not been very articulate about this in these discussions, but everyone who spoke has assumed some major conservation activity. Everyone has estimated energy requirements by the year 2000 as really far down from what they were a few years back. And when I talk about levels of energy of 90 to 150 quads, both those numbers are lower than the numbers that were projected three or four years ago. So the Department of Energy is depending on a tremendous amount of conservation, and it is developing. There is much more that can be done, but many significant things are going on in conservation; the field is so broad it is a little difficult to get into specifics.

From the floor: Can the United States go on, as your forecast would suggest, putting out 40, 50, 60 billion dollars a year to OPEC for many years without a collapse of the international monetary system?

Myers: That is a tough question. The U.S. is increasing its export of dollars each year, and it used to be said that this money was getting re-invested back in this country. But you notice that Saudi Arabia Incorporated is beginning to diversify, beginning to put its money in other places in the world, and it cannot help but have an impact. I have tried for two years in Washington to get some department to give me an inkling as to how much it is worth to produce a million Btu's of energy in this country as compared to importing it. I cannot get an answer. But it is clear to me that the nation will be better off if we ourselves can produce, get out of the import loop.

Chairman: This is a synthetic question made up of a number of fragments. It questions the value to the nation of the Department of Energy. "Attention is called to the fact that the budget of the Department of Energy is roughly equal to the profits of the entire oil industry, and a question is raised: Would the government be better off without having such a department at all?" *(Audience applause)*

Myers: Let us consider nuclear weapons. The Department of Energy builds all the nuclear weapons. Somebody can applaud the idea of getting us out of the nuclear weapon business but it *is* a part of the security of this country today. About a third of [the Department of Energy] budget goes into that. About a third of it goes into R&D. About a billion dollars is spent on fossil energy, including work I have mentioned on liquids plants; about a billion dollars on nuclear activities with about half of that in breeder reactors and half of it in light water reactor activity. And then the rest of it is split up. A very large amount is going into solar energy including photovoltaic activity. Without the DOE, some of those technologies would never get off the ground until the price of oil went up to the place where industry decided it was going to stabilize at the high price and then was willing to go forward.

Take liquids-from-coal for example. DOE is starting a plant now—it will be on line by 1983, '84. Industry, even if the price of oil went to $25, would feel that if the United States started replicating plants of that nature, OPEC might drop the price again so the country could be in the position where it would never quite get over that hump of building coal liquefaction plants. So I think somebody is needed to do this kind of work. It could be another agency, but a budget needs to be involved in R&D that will bring new sources to a place where industry can take over on them.

Whether the Department of Energy is net gain to the country I do not know, but let me say one thing about that. There were four or five separate pieces of the government—the Federal Power Commission, the Federal Energy Agency, the Energy Research and Development Administration and other miscellaneous small parts of [the Department of] Interior and other bureaus—and these were brought together in one operation, to get a little more systematic idea of the views of the regulators and the power operators and the R&D people. The Department has not done a great job yet, but we are working on it and I think we are beginning to have some cross-talk within the Department. There are actually a few regulators sitting down with some R&D people and dealing with the problems. And that, I think, is where we will come out ahead as a Department of Energy.

I agree that the Department of Energy, with a 10 billion dollar total budget—eighth in the government in personnel, tenth in the government in expenditures—is very large, but I could think of a few other large departments that people could applaud getting rid of too.

Chairman: Here again, a large group of questions having to do with relating productivity growth with the problems of energy availability and energy conservation. To sum up, you mentioned that the U.S. appears to be conserving energy because energy as a fraction of the GNP is going down. Are we really conserving energy or are we merely substituting labor for energy, thus, in effect, decreasing the standard of living?

Myers: No, I think energy is actually being conserved. The simplest example for all of us to see is the increase in gasoline mileage of automobiles; that is a dramatic thing and it will be more dramatic in the future. Or the things that industry are doing—where given a certain automatic line, or automatic production operation, through just good ingenuity they find ways to [operate] with less energy: recuperating heat, reusing heat from the end of a heat-generating system; those are the kinds of things industry is beginning to do. Insulation of houses—just the general air tightness of a house becomes important. People are beginning to be concerned about that in their residences. So there really is energy saving going on now.

I think of one beautiful example. While I was in Washington at NASA, in 1974 the temperature of our building was being balanced by bucking the heat system against the air conditioning system. Energy was so cheap then that that is the way the temperature was balanced. And that was not just government buildings; that was the standard in Washington, D.C. at that time. That is not done any more. That is energy saving.

Chairman: One more question, a general one this time, about technological innovation in the nation. Attention has been called to certain projects, for example the Barstow Solar Tower, where hundreds of millions of dollars are spent. The question is, in this nation, with so many technological innovators, it is startling to some to see a lack of proposals for R&D that are really radically innovative. Why is that?

Myers: A good question. The Department of Energy has an inventions program where people can submit proposals which are then gone through with a very thorough peer review, looking for the good ones. DOE supports individual inventors, and the inventors end up with all the patent rights. That is a good deal for an inventor who has an original idea in energy. But the Department is not seeing bright brand-new ideas that will make any major contribution.

I think the summary of this whole energy matter is that it will take all the innovativeness and all the invention we can get in this country to solve our energy problem. It will take all the energy sources America can come up with to offset this terrific flow of oil that is depleting the world supply. And I do not know what is missing. Is it university programs in energy? Is it a program that stimulates through prizes? DOE is trying now to develop broader university research programs in fossil fuel, broader programs in solar energy and new energy sources. It is not easy to see what will attract the brains that will give good answers, but it is clear that it has to be done.

John S. Foster, Jr.

Energy Technology and the Abundant Energy Future

The theme of this conference is refreshing—that the United States will in the future have abundant supplies of energy even though they will cost more than in the past. With this theme, we can focus on ways to assure that future rather than continue to "view with alarm," which does little to point the way to a favorable outcome of our present situation. If we are to have a satisfactory energy future, we must decide on our objectives and what we should be doing now other than what we are already doing, that will lead to that future.

In general, I share the optimism expressed by the theme. I believe the United States can have a bright future if we manage our energy affairs well now. There is also the likelihood, however, of a bleak future if we fail to take appropriate actions.

My agenda is a simple one. First I plan to say a few words about the energy supply and demand outlook, and particularly about the assumptions underlying most current projections. Then I will discuss the risks that are associated with those assumptions and suggest the need to guard against a gloomy future, the need for a hedge to reduce those risks. Next is a proposal for such a hedge. Finally, I will point out that there is a potential silver lining to the energy clouds we see today.

John S. Foster, Jr., *is Vice President, Science and Technology, for TRW Inc., coming to that company from the U.S. Department of Defense where he was Director of Defense Research and Engineering. Earlier Dr. Foster was Director of the Lawrence Livermore Laboratory of the University of California.*

The Energy Outlook

The U.S. energy outlook between now and the year 2000 has been examined many times, with results that are generally in good agreement. Parenthetically, I would like to point out that this agreement is more worrisome than reassuring; it may indicate reluctance to deviate from the crowd rather than reliable data. Without going over the details of the many energy supply and demand forecasts, I think it is safe to say that they generally agree on the following points:

1. The growth rate of the U.S. economy is expected to decline to about 2 percent annually.
2. Even with this decline, we will continue to import about half the oil used to meet our energy demands.
3. World oil production is expected to peak around the year 2000 and decline gradually thereafter.
4. Beyond the year 2000 there will be a gradual shift to inexhaustible energy resources such as fusion, breeder reactors, and solar energy. The year 2020 is often taken as the limit of credible projection, but of course choosing 2020 does not automatically provide results in reliable 20-20 foresight. Uncertainties are very large at that distance in time; in 1960, few would have placed much confidence in a prediction describing our present energy situation.

To make forecasting possible at all in the face of basically intractable equations that would have to include political, economic, social, and religious factors and their spheres of influence, forecasters normally state their simplifying assumptions. Those assumptions are generally along the following lines:

1. Government energy policies will be consistent with national needs.
2. The average growth rate of the real GNP will, as stated earlier, decline from about 4 percent to about 2 percent over the next twenty years.
3. Additional conservation efforts will reduce energy demand by the equivalent of 10 to 15 million barrels of oil per day over the next twenty years.
4. There will be no limitation on our access to imported oil.
5. Environmental obstacles to development of our energy resources will be overcome. (These apply especially to nuclear, coal, and shale sources.)

Despite general concurrence on the supply and demand forecasts mentioned above, they are not very reassuring. In particular, our continued dependence on imported oil should make us uneasy, and there is no clearly evident replacement for the future declining supplies of oil and gas. Looking at the forecasts in more detail, we find that:

By 1990

- total energy use will rise by a third (reflecting a 3 percent GNP growth rate over this period)
- half will be from oil
- half of that oil will be from domestic sources
- half of the domestic oil will come from new discoveries
- 20 percent of our energy will be from natural gas
- half of that 20 percent will come from new discoveries
- 10 percent of our energy will come from nuclear plants, which is three times their present capacity
- 15 percent of our energy will come from coal, nearly doubling the present output
- the remaining 5 percent will come from hydroelectric and other solar sources and geothermal sources.

By 2000

- total energy use will rise by an additional 20 to 25 percent
- 30 percent of total energy will be from oil
- half of that oil will still be imported
- most of the domestic half will be from new discoveries
- 15 percent of our energy will be from natural gas
- most of that gas will be from new discoveries
- 35 percent of our energy will be from coal, requiring a threefold increase in coal production
- 15 percent of our energy will be from nuclear sources
- 5 percent will come from hydroelectric and other solar energy sources, plus geothermal energy.

These are only estimates, of course, but they represent the general consensus. What I want to point out is that should these predictions become reality we will be very fortunate indeed. For one thing, they assume sustained growth of the economy even though the rate is slowed; admittedly one or more recessions could change the picture considerably. Second, they are based on the assumption that the United States can import all the oil it needs; that means that we will have the money to pay for imports, that the producers will be willing to produce the amounts we require, and that there will be freedom of the seas permitting imports over long supply lines. Third, the assumption is we will find significant new domestic sources of oil and gas; obviously we do not know today where they will be or even that we will find them. Fourth, these projections assume that we can burn all this fossil fuel without seriously affecting the earth's heat balance despite the increase in production of carbon dioxide; and fifth, that the growing nuclear contribution can be realized even though there is still serious opposition.

Each of these conditions is possible, perhaps probable, but not certain. Taken in combination, they require a fair amount of luck on our part—in my view, too much luck to plan on. As a consequence, I find the forecasts (and the planned actions on our part) to be unacceptable. One real possibility is that the OPEC nations, especially those in the Middle East, may choose to limit their production as a way of prolonging their supplies and thus slowing the decline after 2000. This seems to be a reasonable policy on their part, but one that could have serious adverse effects on our energy supplies for the next twenty years.

To a large extent, the positive elements of the supply/demand forecasts rest on predictions of:

- Massive conservation efforts
- Very substantial new discoveries of domestic natural gas and oil
- A threefold increase in coal production
- A 15 percent or greater nuclear contribution to our energy supply by the year 2000. With a full-scale effort, this contribution could be as high as 20 or 25 percent, greatly alleviating the fossil fuel supply problem.

Accordingly it is absolutely urgent that every reasonable action be initiated and supported just to achieve the listed energy targets; if any of the predictions were to fail, our energy supplies would fall below our minimum needs. And even if the targets can be reached, we will still be heavily dependent on foreign oil sources. At the end of the projected period we will be importing about 12 million barrels per day, or 30 percent more than today.

Meeting the Shortfall

To assure plentiful energy supply in the future, we must consider the consequences of an appreciable shortfall in U.S. energy supplies and initiate those actions that would reduce or prevent such a shortfall. How is this to be done?

Increased prices will generate additional conservation by reducing demand, but conservation is an exhaustible resource. We can reduce consumption, but we cannot eliminate it. It is energy that fuels the GNP, and if we do not have enough even with our best conservation efforts, there will be serious economic consequences. In 1970, energy costs constituted only 4 percent of the value of our GNP, while at the beginning of 1979 the proportion was 7 percent. It can therefore be argued that increasing energy prices to their present levels has not significantly affected the economy. However, the consequences of energy costs reaching, say, 14 percent of the GNP are not at all clear; all we know with certainty is that at some point we will need an energy price ceiling to protect the economy from decline.

It is also argued that reduction in energy demand (by price or by law) will not significantly affect GNP growth. Specifically, the argument is that in the long run, GNP influences energy demand much more than energy availability influences GNP. Examples are given of other countries with lower energy/GNP ratios than the United States and of significant conservation opportunities for vehicles, houses and other buildings. The catch is in the phrase "in the long run." Other countries have adapted to their conditions (high gasoline prices for example) over several decades, and their geographies and population patterns favored such adaptations. We are now in the process of adapting, but it is a long process; replacement of our stocks of vehicles and buildings will take much more time than is available to us to react to our current situation.

A substantial reduction in the availability of imported fuels would in the short run have non-linear and possibly catastrophic effects on our economy. As noted, virtually the entire economy is fueled by energy and any reduction of energy supply below our minimum requirements must be reflected in a reduced GNP. Given that energy is currently some 7 percent of the GNP, there is a leverage of about 14 to 1 in dollar terms; a dollar spent on energy generates 14 more dollars in GNP. In the limit, an abrupt reduction of 10 percent in available energy could lead to a drop of up to 10 percent in GNP, which in today's economy is more than 100 billion dollars a year. If the reduction were to continue, it would take many years for the economy to recover.

Finally, it could be argued that a strategic reserve of oil would make up for the lack of increased domestic capacity. Such a reserve does mitigate the effects of a sudden and limited interruption of oil supplies, but a reserve does not impose a price ceiling for imported oil or protect us from prolonged chronic shortfalls resulting from the reluctance of some producers to exploit their oil reserves fully.

In other words, our forecasts assume a dependence on luck to keep us on what is essentially a tightrope controlled by others, and one without a net under it. I think we should start looking for a net, or perhaps it should be called a hedge. We are betting on the successful outcome of multiple risks, and the prudent thing to do is to hedge our bet in case one or more of these outcomes is not successful.

Reducing Our Risks

This conference offers a good opportunity to suggest that we take action to reduce the risk this nation is running of finding itself with no way to meet all its energy needs. Our current long lines of cars waiting for gas in the Los Angeles area provide a slight indication of just how sensitive we are to interruptions in the flow of imported oil; it is slight in terms of effect on our economy, but even hearing about long gas lines is enough to get full public attention. And it seems likely that this tightening supply situation could persist for longer than just a few weeks. That in turn should be enough to get the attention of the executive and legislative branches of government, especially just before an election year, and that is where the action must start.

What we need as a hedge against the risks I have outlined is a larger assured domestic supply of liquid fuels and gas, one that does not depend on luck or on technological breakthroughs. Nearly all analysts agree that the only such supply is our own deposits of coal and shale. In the era of cheap oil and gas they were too costly to exploit, even though the technology for producing both liquid fuels and gas from coal was established and in use in other countries. The process for extracting oil from shale was also known, although both the cost and a number of U.S. environmental problems have prevented any large-scale development. We even have large reserves of gas in the so-called "unconventional" sources—tight sands, shales, coal seams, and geopressured formations—but the price, and in some cases the technology are not well known.

The Potential of Synfuels

Liquid and gas fuels produced from coal, shale, and biomass are collectively known as synthetic fuels—synfuels for short. Many studies and economic analyses of synfuels have been made in the past five years, and several new and apparently more efficient processes for coal conversion have been brought to the development stage. The Lurgi process for coal conversion, however, has been in use on a large scale for decades and its economics are well known. Technologically it is a low-risk process, but cost, environmental and licensing questions prevent the building of Lurgi plants. We have the beginnings of a contribution from biomass in the form of the gasohol now being marketed, and no new technology is required to double the small percentage of our energy now being supplied by burning wood.

The rising price of oil is rapidly eliminating the cost barrier. There are at present two firm proposals for building privately-financed coal conversion plants, awaiting only the necessary permits. The general picture of synfuel economics is given in Figure 1, which indicates the ranges of synfuel costs given in current estimates and compares them to two different scenarios for world oil prices.

Fig. 1. Petroleum and Synfuel Price Patterns

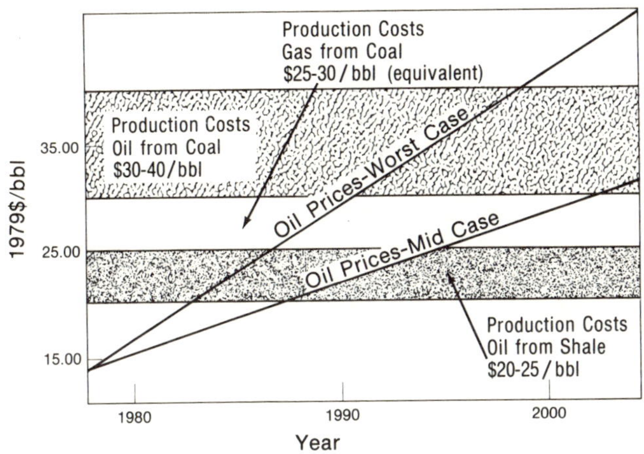

Production costs for synfuels should not in general increase in constant dollars, while world oil prices, in response to market forces, will. The implication is that a substantial synfuels production can clamp a ceiling on the average price we have to pay for oil and imported gas. The ceiling, of course, is only effective if the synthetic capacity developed is equivalent to a substantial fraction of U.S. oil imports. So, an effective synthetic fuel program must have the capability for both providing a floor on domestic supplies of oil and gas and providing a ceiling on prices of imported oil.

It is possible, but not at present likely, that the world price of oil would go down in response to competition from synfuels and make our synfuel capacity uneconomic (unless the synfuel industry were protected by tariffs). It is certainly not going to come down in the absence of such competition, and that is why we need a synfuels capacity, as a hedge against all the uncertainties associated with imported oil. If the capacity turns out at some point not to be needed, it will have served as cheap insurance.

All this has been pointed out before, but despite attempts to get a synfuels program started, nothing has happened. The oil embargo of 1973-74 did not do it, the Iranian crisis in early 1979 did not do it, and the recent gasoline shortage has not done it yet. Meanwhile, we are running out of time; it takes time to build a capacity and if we do not start now we will be at the mercy of events beyond our control and running the risk of a serious economic dislocation that would cost many times the price of a synfuels program.

Needed: A Government–Industry Partnership

What is the problem? In my view, the problem has been the lack of a coordinated program undertaken jointly by government (including both energy and environmental interests) and those segments of industry that have the capability to design, build and operate the plants. Certainly it is the responsibility of the government to provide for the general welfare, as stipulated in the Preamble to the Constitution, and the general welfare is strongly dependent on adequate supplies of energy. Given the present environment of the marketplace, only the government can take the steps necessary to change that environment and generate the economic force to bring industry into the picture, to reverse the current trend toward a potentially disastrous situation. And only industry can do the job.

The Executive Branch has to take the lead by setting specific time-phased goals for synfuels production, enlisting the cooperation of industry in making a rational plan, and reconciling environmental interests with energy requirements. When a plan is developed, the Congress must approve and authorize the program and appropriate the funds needed to carry it out.

Specifically, I suggest that the President might request the Secretary of Energy to prepare, within six months, a plan for putting into place over the next 20 years a domestic synfuels capacity amounting to the equivalent of 7.5 million barrels of oil per day. The plan would be developed jointly by the Department of Energy, the Environmental Protection Agency, and industrial firms capable of managing the design, construction, and operation of the plants. It should be the result of negotiations among these parties, and not a government program offered to industry on a take-it-or-leave-it basis.

My suggestion is that the plan call for the immediate construction of ten initial plants, each with a daily capacity equivalent to 20,000 barrels of oil. The plants would be of different types, using coal, shale, and biomass, and producing both synthetic liquid fuel and gas. While the different sources and products require different processes, most should be based on processes with low technical risk. A small number, perhaps two or three, could be selected that were judged to have moderate risk but with the promise of lower product cost.

Synfuels–Phases and Costs

An approach to designing the synfuel initiative might consist of three phases:

- Pioneer Phase, now through 1985—development and operation of pioneer plants generally based on second generation synfuels technologies
- First Production Facilities, 1984-1990—construction and operation of first production facilities based on pioneer plant experience
- Later Production Facilities, 1990-2000—replication of first production facilities to achieve target production rates of 5 million barrels per day in 1995 and 7.5 million barrels per day in 2000.

Table 1 shows the number of plants and production levels for each phase for 1985, 1990, 1995 and 2000. Assumptions underlying Table 1 include the following:

- Pioneer plants can be built and operated in time to permit decisions to enter the First Production Facilities Phase about 1984 with second generation technology.
- First Production Facilities can be designed, built and placed in operation in 5-6 years. Later Production Facilities can be constructed in 4-5 years.

This program would include plants producing liquids and gas from shale, coal, and biomass. The particular mix of plants would be based on the success of the various processes and on the evolution of demands for liquids and gases, especially in the industrial and residential/ commercial sectors.

Such a program offers several opportunities for modifying product

Table 1: Plants and Production Rates in Synfuels Initiative
(Capacity in Millions of Barrels of Oil per Day Equivalent)

	1985		1990		1995		2000	
	Plants	Cap.	Plants	Cap.	Plants	Cap.	Plants	Cap.
Pioneer Phase*	**10**	**.2**	—	—	—	—	—	—
First Production Facilities	—	—	**20**	**1.0**	**20**	**1.0**	**20**	**1.0**
Later Production	—	—	—	—	**80**	**4.0**	**130**	**6.5**
Total	**10**	**.2**	**20**	**1.0**	**100**	**5.0**	**150**	**7.5**

*The pioneer plants are each of about 20,000 barrels per day capacity; other plants are assumed to have a capacity of 50,000 barrels per day.

mixes, scaling plans down, or deciding not to enter subsequent phases. The first such opportunity would come about 1985 when the First Production Facilities Phase is about to begin. At this point, about $5 billion would have been invested in the pioneer plants. A second opportunity is presented about 1990 when the Later Production Facilities Phase begins. At this time the ultimate scale of the initiative could be reduced or the initiative terminated if conditions so indicate.

It seems clear that implementation of a synfuels program like that outlined here will require substantial government support. The Pioneer Phase alone might cost five billion dollars. More important, this phase is inherently risky. Subsequent plants, while based on better technical and economic data, are much larger. Their scope and cost may very well be too great for any but the largest companies, and it is very doubtful that a synfuels industry will materialize without extensive federal financial and environmental support. Perseverance and consistency on the part of the federal government will also be needed.

If we go ahead with the entire program, it will cost on the order of $200 billion over twenty years. This is the equivalent of a 7 to 10 cents per gallon tax on gasoline for that period; (our gasoline tax now is the lowest among industrialized nations). The cost is about a half percent drop in GNP over the 20-year period—and that may be the least painful of the alternatives. (All these figures are in 1979 dollars.)

How to Do It

There are several well-known methods by which the government can use its financial or legislative power to carry out a program. The choice of a particular method will be made in the course of developing the plan, but the choice of that method is relatively unimportant in relation to the importance of agreeing on the objectives, the plan, and carrying it out. Some of the methods that have been researched in depth and proposed are the following:

1. The government mandates a certain increasing percentage of synthetic fuels that must be included in each supplier's total output. The cost is automatically passed along to users.
2. The government guarantees a price for the output that will justify building and operating the plants. If the market price is less, the government pays the difference. The guarantee must be for a long enough period to provide sufficient confidence for industry to make the initial investment.
3. The Department of Defense, a large fuel user, is directed to buy a stated quantity of synfuels. Again, quantity and price are sufficient to justify building the plants.
4. The Congress creates a government corporation that will enter into joint ventures with industrial companies to build and operate the plants. The government corporation recovers its investment only if the venture is profitable.

While each of the above has advantages and disadvantages, I have a concern that proponents of one or another of these methods will feel that choosing a particular method is more important than getting started, and that the patient will die while the surgeons are arguing about which scalpel to use. Our hope should be that through negotiations between the government officials and the performing industry some mutually agreeable set of methods will be found that the Congress would endorse. There must be a sense of urgency great enough that even fervently held approaches can be subordinated to getting the job done. Our future viability as a country may well be at stake.

The Silver Lining—Inexhaustible Energy Sources

The synfuels program suggested here is not a long-term solution to the problem of meeting our energy needs. It is not an exciting solution; it is simply the only one we have that will protect us while we find long-term solutions.

Nearly all analysts agree that in the long run the United States, and the rest of the world, will have to turn to "inexhaustible" sources of energy, namely, solar and nuclear energy. The program outlined above will have made a start with the synfuels from biomass plants—a form of solar energy. The other forms are well known and are the subjects of active Department of Energy programs. Fusion is most often mentioned as an inexhaustible nuclear source, but it may be worth mentioning that nuclear fission can be extended far beyond its currently projected useful period of energy production by one or more of the following technologies:

- Use of lower-grade ores
- The fast breeder reactor
- The thorium cycle
- The fission-fusion hybrid.

Fusion

I will not discuss these technologies further here, but I would like to say just a few words about fusion energy.

The concept is simple and proven. With the proper conditions of temperature and pressure one can fuse together rare forms of hydrogen (deuterium and tritium) and create and release large amounts of energy. This process has been demonstrated in thermonuclear explosions, but to apply the process to commercial power generation requires a very different technology. The industry has been working on the problem for a long time now, but achieving controlled fusion has proved to be harder than expected. Research scientists are still waiting for the net energy breakthrough, when more energy is obtained from the reaction than was used to create it.

It is finally beginning to look as though that point may be near. About a dozen teams of researchers, using different kinds of machines, have been making steady progress in recent years, both in the inertial and magnetic confinement approaches. There is now an air of expectation, and during the next five years exciting announcements to the effect that they have surpassed the breakeven point may be forthcoming.

Granted these scientific breakthroughs, a critical question is how practical fusion will be as a means of generating electrical power. The process will have to involve not only the scientific laboratories but also teams of people who have successfully engineered practical fission reactors; and so a lot of engineering remains to be done after the scientists have shown that it works. A number of preliminary concepts have already been developed, and I find no real "show-stoppers" among them. A lot of hard work, good luck and money will be needed before we start feeding fusion power into our electrical grids.

Action Now

My point is that there is reason for optimism regarding our energy future beyond the year 2000. First we must get to the year 2000, however, without intervening economic problems that can keep us from pushing hard on all the technologies that offer promise to deliver major amounts of energy in the near term, the medium term, and the long term. That requires action now.

In recent years we have experienced a period of more or less public disenchantment with technology as a way of solving our problems. Without an aggressive technological effort by the United States there is little hope for an era of plentiful energy. With such an effort we can not only provide plentiful energy, but in the process of properly applying science and technology to one of our most pressing needs we can also help to restore public confidence in science and technology and provide for a better life.

Discussion

From the floor: Would you be willing to add reprocessing and breeders to your list of high confidence and, presumably, high priority mechanisms?

Foster: Absolutely yes. Let me expand on it, if I may. The difficulty at the moment lies not only with making sure light water reactors are designed, constructed, and operated properly and safely. Those are things that can be done. But it is necessary to reckon with the fact that exploration for uranium has been rather severely limited, and so in the years ahead, that can be a pacing element in the supply of fuel for the reactors. In the longer term, a consideration is that these reactors are using the fuel about a factor of 50 or 60 less efficiently than they could and should. It is not possible to take advantage of the possibility of breeders without reprocessing. So the whole matter of reprocessing and waste disposal and breeders is crucial if there is to be nuclear energy in the 21st century and beyond. In my mind there is just no question whatsoever.

I would also note one additional promising development that could come from the fusion program. At the moment, as I indicated, the researchers are quite confident that they will achieve what is called "breakeven"—a thermonuclear burn. That can be by either the magnetic or the inertial implosion approach. An important thing to note about the fusion process, however, is that for each reaction one can get the equivalent of about 4½ neutrons, and it is neutrons that do the breeding. The present breeder system gives about half a neutron. What that means is that with the fusion process, even though it works far less well than people hoped, there is the possibility of making it into a breeder—a breeder that would provide fuel to support reactors in far larger numbers than the current liquid metal system. So fusion is important not just to produce fusion power but to produce fission power.

From the floor: You seem to prefer higher Btu gas from coal gasification. What is your reason?

Foster: I have a preference for it primarily because a distribution system exists. If pipeline-quality gas can be obtained, this is a much more flexible fuel.

From the floor: It occurs to me that in the field of synfuels the price cannot really be known until a plant has been built and operated. There are considerable uncertainties, perhaps a factor of two. Also, whether the plant produces 20,000 or 50,000 barrels a day equivalent, that performance could be a lot better or a lot worse, and one process work better or another not so well.

Foster: I completely agree. That is why I made the point of trying for, say, ten plants at the outset. We in industry know it can be done with older generation systems, and the question will be, should we try a second generation? Should we start pushing a little faster a third generation approach? I completely agree that there is uncertainty and a need to start a number of different plants.

From the floor: I have heard about two interesting proposals for new sources of oil, and I wonder if anyone present has additional information about them. One concerns pouring polymer down into long-dead wells, from which it may then be possible to extract five to six times more oil than was ever found before. The other is from Nobel Prize winner Melvin Calvin of the University of California, Berkeley, who proposes the growing of a hydrocarbon-rich plant from the euphorbia family as a source of oil. He estimates the crop would yield to the farmer (this was two years ago, it may be a considerably higher figure now) $2,000 an acre. In a certain climate, he could have three harvests a year. Furthermore, this could be grown in the desert and would not be competitive with agricultural lands.

Foster: I have talked with Mel Calvin about his idea for growing the euphorbia plant in the desert and extracting the oil from it for processing into fuel. It is an exciting idea, and some initial effort is now underway.

John Swearingen (from the floor): Many people in the oil business have been doing research on the other idea for twenty to thirty years; our own company has some operations in the field in West Texas today to try to determine the economics of this kind of project.

The principle on which it works is that by introducing a polymer material or any other type material (caustic soda would do this too), one can alter the surface tension relationship between oil and the rock in which it is enclosed, and by doing so make it more available. But even if the oil is loosened up, it will not flow out of the well, any more than gasoline flows back up out of the tailpipe of an automobile. Something has to push it. And the only way it can be pushed is by injection of gas or water or some other fluid. The economics of this are not established at present. The biggest question is, how much loss of the chemical material takes place underground? This is not like a concrete sidewalk; it is not uniform. What the loss of chemicals will be in relation to the oil that can be produced is still to be determined. Much work is being done.

From the floor: You have talked about the centralized concept of having a government/industry relationship for building a synfuels industry in the next couple of decades. What is your opinion about a partnership for a *de*centralized option, such as solar systems in factories, shopping centers, perhaps even for entire cities? What is the feasibility of that kind of partnership, that kind of mix?

Foster: I am all for it. The United States needs every bit of energy it can get. We could hope that in the decentralized system the questions regarding capital and the bureaucracy would be far less and one could move more rapidly. That is not always the case. But it was possible for TRW, for instance, to put a solar system on the top of one of our buildings in Houston; we had no difficulty with the federal government or the state and local officials, and we are saving energy. I would think others could do the same thing. It would be particularly good for people with pools, to heat them with solar instead of firing them on gas.

Charles J. Hitch

The Price of Energy—and Prosperity

Like the rest of the invited participants, I have been asked to examine the hypothesis that "America can prosper in an era of plentiful, although higher priced, energy."

An interesting proposition, and one, if sufficient emphasis is placed on "can," I am inclined to answer in the affirmative. But when I am asked an all-encompassing question like that, my immediate reaction—honed by years of dueling in the groves of academe—is to fire back with a few questions of my own: "What is meant by (a) "can prosper," (2) "in an era," (3) "plentiful," and (4) "higher priced?"

Energy in Relation to the Economy

In a few minutes I will address each of these questions in presenting my response to the hypothesis, but to do so I first need a bit of context. Let me begin with what I know (or believe) about relations between energy and the economy. I summarize the relations in the six following propositions:

1. The size of the economy as measured, for example, by GNP, strongly influences the demand for energy. Other things being equal,

---

Economist, educator, and President Emeritus of the University of California, **Charles J. Hitch** is a participating guest in research in the energy and environment division of the University's Lawrence Berkeley Laboratory. He has served as Assistant Secretary of Defense, has been adviser to ERDA and the Gas Research Institute, and is a member of the National Petroleum Council. Dr. Hitch was President of Resources for the Future at the time of this conference.

including energy prices, a 10 percent increase in GNP increases energy demand by about the same percentage. Other things are seldom equal, however, so the ratio of energy to GNP often changes over time and differs from region to region. It can vary with life styles, industrial structure, and climatic and geographical factors, and is strongly influenced over the long run by the price of energy. In fact, energy use per dollar of GNP varies greatly (by as much as a factor of 2) from one industrial country to another, and by even larger factors among the states of the United States. We are doing a lot of fascinating work at Resources for the Future, measuring and accounting for these differences.

2. Conversely, the effect of energy availability or energy prices on the economy, in the long run and within wide ranges, is *much* less than GNP's influence on energy. There are two reasons for this lesser effect.

Energy's Cost Drag

First, the cost of supplying primary energy is small in relation to GNP (typically, in the past, 4 to 5 percent, and even now, after all the OPEC boosts, only about 7 percent). So even a further doubling, say, of the real cost of supplying primary energy has a much smaller impact (at most a reduction of 7 percent over the period of the doubling) on GNP. If we should meet again in 2010, thirty years from now, and energy costs and prices have only doubled over that period, we would find that the effect on productivity of the higher energy costs would have been a maximum of one-fourth of 1 percent per annum. This is not a trivial drag on the growth of GNP—indeed, it would amount to many billions of dollars annually—but it certainly would be consistent with "prosperity" if we can return to anything approaching past norms of productivity increases, that is, on the order of 2 to 3 percent per annum.

I think this range of productivity increases is possible, but it may not be realized. If the disturbing trends of the last five years continue and productivity growth declines to, say, one-half of 1 percent, then the energy cost drag becomes relatively much larger. It would mean that we would be using up what little flexibility and margin we had in the economy, bringing ourselves perilously close to the point where U.S. society as a whole would be locked in a vast zero-sum game; no-one could gain without someone else losing the same amount. As the British are finding out, this is a recipe for social tension, confrontation, unrest, and inflation. Such a dismal state of affairs is not likely, but we should be very clear that it remains a not-too-distant possibility should productivity—for reasons having little or nothing to do with energy—slump further.

Energy's Drag on Productivity

The second reason why energy's effect on GNP is much less than vice versa is more complex. Let me emphasize first that one-fourth of 1 percent per year is energy's maximum drag on the economy. It assumes that there is no elasticity in the demand for energy, no possibility of substituting other factors or goods and services for energy as its price rises, and that we must therefore divert resources to expand energy supply by brute force.

We *know* this is wrong. We know there are many ways of substituting capital and labor for energy as energy becomes more costly. We know we can make our houses more energy-efficient by insulation and simple changes in design. We know we can make our cars more energy-efficient. There are similar possibilities throughout commerce and industry. And consumers can buy fewer energy-intensive goods and services as their relative prices rise.

We do not know the value of the elasticity of substitution of other factors for primary energy with any precision or confidence; but if it is at least −0.3 (i.e., to a first appropriation, if the demand for energy falls by at least 3 percent when its price increases by 10 percent), the one-fourth of 1 percent annual drag on productivity calculated above falls to something very substantially lower.

Let me illustrate what I mean by 7 percent being the maximum loss by a homely example based on beef. Suppose that a beef-loving American (or perhaps, more plausibly, Argentinian) consumer, spends at present prices 7 percent of his income on beef. Now suppose that the price of beef doubles. The worst that can happen to him is that his cost of living will increase by 7 percent, and his standard of living fall *pari passu* by 7 percent, corresponding to our worst energy case above. And this will in fact happen for the rare beefaholic who stubbornly insists on consuming just as much beef of the same quality as he consumed before.

At the other extreme there will be some consumers who, at the original price, were near the margin of indifference between consuming beef, on the one hand, and lamb, pork, and chicken, on the other. When the beef price doubles they will switch, at a trivial loss in their standard of living. The vast majority of consumers will be inbetween; confronted by higher beef prices they will adjust by making some substitution, continuing to eat beef, but fewer times each week, and perhaps cheaper, less desirable forms like hamburger (low-grade hamburger is the coal of the beef market). They will experience some loss in their standard of living, but much less than 7 percent.

3. While the economy can adapt to scarcer and more costly energy in the long run, it cannot do so in the short. An effective embargo, or the interruption of oil imports by wars or revolutions, for example, unless offset from stockpiles, could have an horrendous impact on the nation's production and consumption. Whether rationing was achieved by price or fiat, the short-term percentage reduction in GNP could approximate the percentage reduction in energy supply.* And I should stress that with our domestic oil, coal, and natural gas industries we are comparatively well off in such a short-term pinch. In Japan, for example, there is practically no domestic energy industry, leaving the Japanese far more vulnerable that we are. An interruption of all imports would quickly bring the whole Japanese economy to a grinding halt.

4. How long is long and how short is short? This depends on how *quickly* labor and capital can be substituted for energy. And while some conservation (e.g., turning down thermostats) can be achieved almost immediately, much requires replacing the capital stock of cars and trucks, houses and commercial buildings, and industrial equipment with more energy-efficient forms, which takes years or, in some cases, decades. A recent estimate,** consistent with U.S. experience since the quadrupling of imported oil prices in 1973-74, is that only about 10 percent of the potential savings can be achieved in the first year, and 10 percent of the remaining potential in each subsequent year. Even conservation takes a lot of time.

5. What are the implications of all this for the "need" for energy thirty years from now? We can't say precisely, because it depends on how rapid economic growth will be (quite apart from any influence energy availability may have on economic growth), on how readily labor and capital can be substituted for energy, and on the course of energy prices between now and then. We do not *know* any of these quantities. But if energy prices approximately double, the demand for energy in 2010

*Depending, to some extent, on allocation policies. Discrimination against household in favor of industrial uses could mitigate the effect on GNP, although not necessarily on economic welfare.

**By William Hogan. See his presentation in this volume.

will be less than many think—closer to 100 quads than to 200 (as compared with 75 now), despite continuing economic growth. This makes the long-run energy supply problem look manageable.

6. Manageable, yes, but only if we permit prices to reflect the true costs and scarcity both of energy in general, and of different sources of energy. Consumers of energy will find ways to conserve, if, but only if, they are getting the right signals through prices. If we keep prices low as costs increase and fossil fuels are depleted, we can expect a number of rather unpleasant consequences:

- Demand will increase more than it needs to.
- Domestic supply will increase less than it would with higher prices.
- If *no* limits are imposed on imported oil, imports will expand to fill the gap caused by price controls.
- If limits *are* imposed on imports, energy will have to be rationed.

The costs of the last two alternatives could be great. An increase in oil imports would drive up the world price, imposing even more severe costs on our friends abroad and on the non-OPEC Third World. Internal rationing—in peacetime, of indefinite duration—would impose a regulatory burden on government that it is incapable of carrying, as well as raising divisive issues of equity of one group versus equity of another, and of equity versus efficiency in the body politic. Even more serious, by keeping prices low we would be failing to provide the incentives to develop and deploy alternative energy sources—renewable or inexhaustible—to replace depleting sources of fossil fuels. We would be replacing a self-correcting mechanism with one that would result in progressive deterioration.

The Goal—Prices at Incremental Costs

It is not easy to get prices up to incremental costs, for all sorts of reasons. The most basic is that no-one likes to pay higher prices for anything, and will use the political process to keep from doing so. At long last we are moving in the right direction with natural gas prices, but ever-so-gradually and uncertainly, after years of underpricing and wasting our cleanest and most convenient fuel. We are still holding domestic oil prices below world prices and thereby subsidizing insecure imports from the Middle East. President Carter now is attempting to break the logjam with phased price decontrol by executive action coupled with a windfall profits tax accomplished through legislative action. Whether this attempt will succeed—and at what political cost—is anybody's guess. About all that is sure is that the battle will be highly emotional. Thanks to our utility regulatory mores we are charging much less than incremental costs for electricity in many regions, in some cases less than half. And given continuing inflation, it is hard to see any politically feasible way of getting out of that bind.

By "regulatory mores" I refer to the almost universal criterion used by utility regulatory commissions: fair return on historical capital costs. This means that when new higher cost plants are built, the higher costs are "rolled in;" i.e., averaged with the lower capital costs of earlier, cheaper plants. As a result, in a period of rising costs—whether real costs or simply inflated money costs—consumers never have to pay, as they should, the incremental costs of new supplies, and accordingly use utility-provided electricity and gas inappropriately and wastefully. As long as inflation continues at present rates or higher, the average price consumers are charged will remain well below the incremental price. Users of fuel oil, by contrast, reap no such windfall. This "wrong signal" could be corrected by the imposition of taxes on utility bills,* but I hear no clamor for that solution from either the utility companies or utility rate payers—indeed from no-one (and this is a fascinating and recent development) except a few activist environmental groups which have taken up "incremental pricing" as a cause. Very logically, too, from their point of view.

Price Relief for the Poor

One step that would help resolve the pricing question—and one that should be taken before next winter—is to insure that poor people no longer die of the cold because they cannot afford minimal heating. As the Wall Street Journal reported in March 1979, shamefully large numbers of people do die every year—most of them poor, most of them residents of the large Northeastern cities—because fuel and utility bills have mounted to a third and in some cases a full half of their monthly incomes. It is hard to fault Congressmen for opposing higher energy prices when some of their constituents quite literally are freezing in the dark.

I have no pet policy to push that will swiftly and easily take care of the problem of poverty and utility bills—that must be worked out through

*Or, less effectively, by some variant of "life line" rate setting.

the political system, and appropriately so. In general, however, I would suggest that this not be defined as an energy problem, but as a welfare problem; it should not be resolved through amendments to energy policy but to welfare policy. Again, energy affects us all, the many who are warm and affluent as well as those who are cold and poor. The former group does not need relief; it needs to learn to value energy more highly.

"Can America Prosper. . .?"

So much for my eternal verities on energy and the economy. They are certainly not all pleasant truths and it is easy to see why politicians have such an awkward time in recognizing them, let alone getting cozy with them. But there they are; they will not go away or become any less painful for being ignored. Indeed, quite the contrary is true; the longer we put off dealing with the tough realities, the harder the impact will be when it is at last unavoidable.

Prosperity and a Time Frame

I want now to return to those questions I mentioned at the outset. Just what is meant by (1) "can prosper," (2) "in an era," (3) "plentiful," and (4) "higher priced?"

I am sure that what is meant by "can prosper" is that the economy can continue to grow, employing the labor force with rising incomes (although not necessarily rising quite as fast as historically or as if energy remained cheap). It does not mean *will* prosper regardless of what happens outside energy. We can all dream up scenarios that would stop or reverse economic growth, but have little to do with energy availability; e.g., a thermonuclear war that destroyed our producers as well as our productive facilities; gross mismanagement of the economy leading to hyper-inflation; or the complete disintegration, in the next generation, of our work ethic. The latter—acceptance of the "less is more" philosophy—might spell prosperity to some Americans, but not, I am sure, to all.

"In an era" I believe means essentially what economists call "the long run." The question is not directed at transitional problems—those that might be caused, for example, by a new embargo or by revolution or war in the Middle East. What is envisioned is a new plateau we somehow reach, in which energy prices are higher but more or less stable. I will discuss later the realism and appropriateness of this vision.

How Plentiful, How High the Price?

The word "plentiful," as used in the title of this conference, connotes two things to me. The first is that there are workable markets in which any individual buyer of energy can purchase as much as he wants at the prevailing price, just as beef would be purchased in the current market. The second connotation is that supplies are secure—that they cannot be cut off by embargoes, wars, and revolutions in faraway places.

And finally, "higher priced." How much higher does that mean? I do not know, and here I think the participants have an obligation to be specific. The hypothesis could well be true at some "higher" level, but not at some still higher one. My best guess is that it would be true up to at least a doubling of primary energy prices. This is all a little imprecise. There is no one price of primary energy, but many and very diverse prices. But if we take the current price of approximately $15 per barrel of imported oil as representative, $30 per barrel at constant 1979 prices would be tolerable—in the long run after the capital stock had had time to adjust.

Cautions

And if a doubling to $30 per barrel is tolerable, what about a further doubling, to $60? Several things begin to worry me as we move beyond the original doubling. First is that wise old general caution: do not extrapolate too far beyond the boundaries of your observed data. More specifically:

a. The very first step in our argument that energy costs have a relatively small effect on the economy was that the cost of primary energy is a small proportion of GNP. With every successive doubling (assuming, as is almost certainly true, that the relevant elasticity of substitution is less than −1.0) it becomes a larger proportion of GNP, and the argument correspondingly less cogent.

b. It is at least plausible that the elasticity of substitution is not a constant, but declines (in numerical value) as prices rise. The easiest conservation comes first. Then it becomes harder and harder as we move from cutting fat to cutting muscle. If the elasticity fell from −0.3 (first doubling) to −0.1 (second doubling), the second would be a disaster.

So maybe, and let us hope, the first doubling is enough. This brings me to the final question I want to ask about the hypothesis on which this conference is based. If it is true (at least up to a doubling of present real costs and prices), is it interesting, is it important, is it useful in pointing out preferred directions for policy in the real world? My answer is a fairly (if not completely) confident affirmative.

Coal, Uranium, and a Cost Ceiling

Is a mere doubling of real energy costs all that we need to adjust to in the long run? Probably, and at worst not much more, for quite a long time.

A likely cost ceiling is provided by our enormous reserves of coal and our uncertain but highly stretchable reserves of uranium. We will encounter some increasing costs as we expand coal production, but I do not see real costs more than doubling, even if the costs of controlling effluents rise substantially; and we are now controlling them so inefficiently that they could well fall if we adopted more cost-effective regulatory methods. Similarly, costs of uranium could rise as we deplete the richest reserves. But the cost of uranium is such a small proportion of nuclear power costs (now about 10 percent at the bus bar) even *much* higher costs would fall far short of doubling the cost of electricity. And if there is real concern about "running out" of uranium there are all sorts of ways of stretching the supply—including cheaper and more efficient enrichment, advanced converter reactors, breeders, and the use of thorium.

So much for electricity; what about our needs for liquids and gases, and in particular, for liquids for transportation? Here again I think coal provides a ceiling. The next-generation technologies for producing synthetic liquids and gases from coal promise to get the costs down to around $30 per barrel (constant dollars). And there are all sorts of other possibilities for getting liquids and gases at the same or even lower prices—new fields as yet undiscovered (witness Mexico); tertiary recovery from old fields; unconventional sources of natural gas; heavy oils; shales and tar sands, to mention a few, all with very large resource bases. And quite probably, within thirty years, some $30 per barrel equivalent solar and biomass sources will come on line.

Of course, all this is not quite certain. There are some very big "ifs" lurking around. Post-Harrisburg fears and rising costs may sharply limit the nuclear options. And carbon dioxide, with its potentially disastrous climatic effects, could conceivably, eventually, do the same for all the fossil fuels. If both occur, we could be in deep trouble indeed. At the very least, we would have to look seriously at the second doubling of costs to accommodate to solar and biomass.

Policy Implications

But probably, for some time, one doubling will be enough. And we can live and "prosper" with one doubling—if Iran and her OPEC partners permit us to live through the transition and get to that plateau about which I spoke. This is important to know, and has highly significant policy implications. The foremost is that we place a high premium—far higher than we have in the past—on *security of supply*. Mere higher costs, at least up to a doubling, and allowing time for adjustment, are not a serious threat to our prosperity, but sudden deprivations are. The case is overwhelming for getting domestic oil prices up to world prices, and getting rid of the cumbersome "entitlements" program which subsidizes oil imports. So is the case for building up a large reserve stockpile of oil. The case is strong, if not overwhelming, for going further and taxing oil imports.

A second policy implication is that we at least develop those nuclear, synthetics-from-coal, oil-from-shale, and other technologies that promise to provide a long-run ceiling for energy costs at about double present prices. We may not need some (or any) of them, and so may never fully deploy them. Fusion, solar, hot rock geothermal, unconventional sources of natural gas, or something else may come along and bail us out. But we had better have them ready to deploy if necessary. This means, in my opinion, carrying *development* through commercial-size plants.

A third policy implication is that we get on with *conservation* to damp down the quads of energy we need to supply in the long run, and thereby the residual quantities of insecure oil we have to import. In the main this should be cost-effective conservation induced by prices, but there are some cases (e.g., more efficient motor cars) where a regulatory assist makes sense.

We frequently hear the argument that because we subsidize imported oil and, in various ways, nuclear power and, in effect by our regulatory mode, electricity and gas, we should—in fairness—also subsidize solar and biomass and geothermal, etc. Because of the importance of conservation, this solution will not do.

It is indeed desirable to get the ratio of price to incremental cost the same for all energy sources—for reasons of efficiency, not equity—but the way to do that is by *getting rid of the subsidies* that now distort energy supply and use, not by adding still more. Getting all energy prices up to, or at least a lot nearer to, incremental costs, will be of immense help in achieving all three policies implied by the hypothesis.

High energy prices are not the problem, as so many people, both in and out of Congress, think. They are the solution, or at least the unavoidable first step to a whole family of possible solutions.

Discussion

Chairman: With your background, which includes leading an educational institution, how do you suppose the public can be brought to understand and accept the kinds of things you have been telling us? What can be done to accelerate the process? What is really practical?

Hitch: It certainly is not easy. All the instincts of the public are against it. As I said, nobody likes to pay higher prices for anything, and people are quite ready to use the political process to keep from doing so. I think it is particularly difficult now because so many of our institutions, including the executive and legislative branches of the federal government, industry, and the universities, just lack credibility with the public; they are not listening. Maybe it is just not possible to do it, or possible only with strong Presidential leadership. There certainly is a role for universities, for the non-profit organizations, for conferences like this, in developing and making available to anybody who can use them, the facts of the analyses which support what would seem to be the way out of this crisis.

Chairman: A number of suggestions show that some in the audience are very interested in the hedge program that Dr. Foster described—the hedge of synthetic fuels—and they are asking what you think is the best plan for causing an accelerated and proper synthetic fuels program to develop in the United States.

Hitch: I am very sympathetic to the Foster plan, and I was a strong supporter of the Synfuels Bill of two years ago, which came within three votes of passage in the House of Representatives. I am inclined to prefer the procedure by which the federal government announces that it is ready to buy a certain amount of certain kinds of synthetic fuels supplied by industry at prices to be determined in a competitive bidding contest among industries and industry groups.

I like that form because its cost is known. It is in the budget, it is out in the open, it gets scrutinized as it ought to be. I prefer it to the alternative that has been suggested by others of my colleagues on this program, which is for federal government to mandate that by the year 1990 or

1995 a certain percentage of oil and gas must consist of synthetic products, which will be defined by the regulations. I do not like this because the costs are buried, they are within industry, within the private sector, they never come into the budget. This is what we have been doing with our regulations controlling impacts on the environment, and I hate to see it extended to still other fields. But if it were a choice of that or nothing, I would have to go along because it is really important to get this program going one way or another.

Chairman: Your career has involved you with defense and security matters as regards energy. Some among your audience would like an assessment from you of how vulnerable you consider the position of the United States to be during this transition period. For example, do we have to give in to OPEC to the extent that we are apparently doing? That is an international aspect. And there is a domestic aspect of vulnerability as manifested by the near panic evident in Southern California at the fuel pump lines. And finally, political vulnerability—the instability shown by the contest of selfish interest constituencies that are very short term. How optimistic are you about our being able to get through this transition phase?

Hitch: I think we are exceedingly vulnerable during this transition phase; the problem is not being exaggerated at all. And we are vulnerable in all the ways you mention.

We are vulnerable to disruption of supply lines in the Middle East—very vulnerable, whether the disruption comes from internal revolutions and wars or from deliberate action by OPEC. I think we are not terribly vulnerable to selective embargoes; I doubt they can be made to stick effectively; the last one did not. But that is small comfort, because *general* embargoes on Western nations are much more serious; our allies in Western Europe, and particularly Japan, are even more vulnerable than we, and pressure can be brought to bear on us, by pressure on them. In fact, that is the way a good deal of it has been happening.

Chairman: Here are questions having to do with the idea of additional investments; for example, accelerated replacement investment. Related to that, our public utilities are prevented, by a present rather illogical pattern of control of pricing and accounting and replacement, from taking the necessary steps to obtain a higher level of investment against consumption in the United States, a problem with which there are, apparently, fundamental difficulties. Would you care to comment on that?

Hitch: There have been discussions in this conference of the physical capacity of industry to provide the investment, and there have been questions about whether the investments that are needed, such as in synthetic fuels, can be made sufficiently attractive to lenders; I have nothing in particular to add to those discussions.

But there is an overall problem relating to investment as a percentage of total U.S. income—the balance between savings and investment on the one hand, consumption on the other. The proportion of gross national income that has been devoted to investment has been falling. It is now well below that of West Germany and all European countries except Britain; it is far below that of Japan. So it is hard to see how there can be, in the limited saving out of present income, enough funds to finance all the things that ought to be done without running into limits that will probably be reflected in intensified inflation—trying to do too much, to continue our consumption, and add on top of it a lot of investment not now being undertaken.

An effect of this reduction in our savings rate has been a much slower increase in productivity of labor in the economy in the last few years. It is really important that the savings and investment rate somehow be raised, but it is not an easy thing to do. There is a great deal of inertia here.

Chairman: You told us that current U.S. energy needs require something like 7 percent of our GNP. A member of the audience asks, "If we were organized in a reasonably ideal and cooperative manner, what would be the right fraction of our GNP that ought to go into the development of alternative energy sources?"

Hitch: I suppose "alternative energy sources" means not just the far-out energy sources but all the things we have been talking about—synthetics and so on. Certainly less than 7 percent. If the synthetic fuels program John Foster proposed would cost $200 billion over a period of twenty years, that is about 1 percent of GNP. Of course, we all want also to develop some alternatives other than synthetics, so let's increase that up to 3 percent. That is a fairly substantial proportion of total savings in this country at full employment levels.

Chairman: A question now that has to do with the impact of advancing technology on all you have discussed. I do not mean specifically the technology having to do with energy; I mean all the other things that change our lifestyle and our options, including options for the use of energy. I refer, for example, to the fact that at some future time (perhaps technically it is possible today but economically, socially, politically it may take ten years to accomplish it), a conference such as this will not require any major transportation [of participants] across the country. I am thinking, also, of technologies that affect productivity because of an improved ability to optimize the control and motion and utilization of our resources; and even advanced technologies that make it possible to tap the wishes of the market so that commitments to production and to inventories will be more closely in accord with what the market wants. If we put in enough of those, they would have enormous effects on our way of life, just as television has had. Do you, as an integrator, have a feeling of optimism or pessimism about the factor of advancing technology?

Hitch: This is what makes me more optimistic than I may have appeared in my presentation; I allowed very little for new technologies coming in to bail us out. I think that in all sorts of ways and directions we will have technological advances. We can dimly see some of them now (they have been discussed at this conference), most of which look benign, most of which look helpful, most of which will aid us with conservation, some of which will enable us to supply energy at lower costs than anything I suggested in my paper. Of course, there is always the possibility that some will not be benign but will work in the opposite direction.

I repeat: with higher prices and lesser availability of energy, people will be stimulated to look for technologies that will help in solving the problems. So yes, I would classify myself as an optimist in this respect. It is a very important source of optimism for us all.

Sam H. Schurr

Wanted: A National Energy Strategy

The proposition we were asked to examine at this conference is "that America can enjoy prosperity in the period ahead of higher energy prices provided that proper policies and implementations are arranged to ensure a plentiful supply of energy at the assumed higher price levels."

My reply to the first part of that thesis is, "Yes, but." The "yes" means that I believe the underlying conditions permit a positive answer; that is to say, I have some idea as to what future costs might be. The "but" means that supporting public policies will be required, as the second part of the proposition says, and that they will need to be timely. Without timely policy actions, the outlook is decidedly unpromising.

The "yes" is based upon the following two broad possibilities, which appear to be achievable:

1. That with new or improved technologies, energy in the commercial forms required can be made available from domestic resources at

Sam H. Schurr *is Deputy Director of the Energy Study Center at the Electric Power Research Institute. Prior to joining EPRI, he directed energy research at Resources for the Future, Inc., for almost 25 years. He has also served as chief economist of the U.S. Bureau of Mines, and has been a staff member of the National Bureau of Economic Research, the Cowles Commission for Research in Economics, the U.S. Bureau of Labor Statistics, the U.S. Department of State, and the Office of Strategic Services. He is the author or co-author of several major books and numerous articles on energy economics and related subjects, and is immediate past president of the International Association of Energy Economists.*

costs no more than about twice today's prices; and that new energy technologies will, over the long run, substantially dampen cost increases compared to the technologies they replace.

2. That in response to economic, technical, and policy forces already at work, energy consumption will grow at a slower rate in relationship to national output than in the 1950s and 1960s; but this estimate needs to be coupled with the proviso that energy consumption in productive activities should not be restricted to levels below those justified on cost-based economic calculations.

In explaining these possibilities, I will draw upon the findings of a recently completed study* by Resources for the Future, Inc. (RFF) with which I was associated before my recent move to the Electric Power Research Institute.

Domestic Energy Resources

The starting point for supplies and costs is the domestic resource base for mineral fuels. Although imports, particularly of oil, will continue to be an important factor in domestic supply for an extended period, they cannot be counted on. Experience indicates that prudent plans for the future must be based mainly upon domestic energy resources.

A conservative estimate of the combined domestic resource total for all mineral fuels used in today's commercial technologies—oil, gas, coal, and uranium—ranges from about 40,000 to 110,000 quads,** depending on whether the energy content of uranium is measured with or without "breeding," compared with current annual consumption of about 75 quads. If all energy sources were readily interchangeable and there were no concerns about public or environmental health, there would clearly be no danger for a long time of running out of domestic resources to satisfy future U.S. consumption (although there could be problems in gearing up production in a timely fashion).

---

*The study was financed by a grant from The Andrew W. Mellon Foundation and is published as *Energy in America's Future: The Choices Before Us,* a study by the RFF National Energy Strategies Project (Sam H. Schurr, director, Joel Darmstadter, Harry Perry, William Ramsay, Milton Russell) (Baltimore, Maryland: Johns Hopkins University Press for Resources for the Future, Inc., 1979).

These remarks represent my interpretation of the findings of the study. They should not be attributed either to Resources for the Future or to my present employer, the Electric Power Research Institute.

**1 quad = 10^{15} Btu; this is roughly the equivalent energy content released by burning one-half million barrels of oil per day for one year.

Assuming, for instance, that energy consumption grew at a 2 percent rate annually (a somewhat higher rate than we estimate), the smaller total would still "last" for about 125 years, and the larger (with nuclear breeders) for about 165 years. Because, in the long run, other factors would limit such steady compounding of consumption growth, and much lower grades of resources could become economic (especially for uranium used in the nuclear breeder reactor), even longer periods of sufficiency are likely.

However, such a conclusion holds only for energy resources in the aggregate, and the aggregate is obviously dominated by coal and uranium. Both are useful today principally in generation of electric energy, which makes up about 30 percent of the total consumption of fuels. Of the remaining 70 percent, more than nine-tenths is consumed in the form of oil and gas; and the natural resources of petroleum and natural gas present an altogether different picture. To illustrate: even if imports of oil were maintained at the 1976 level of about 40 percent, conventional domestic oil and natural gas resources, on a conservative reckoning, would last not much beyond the year 2000 if oil and gas consumption grew at an annual rate of 2 percent.

Thus, even if coal and uranium resources together offer long-term supply assurance, this will be realized only if we can assume the successful development of reasonably benign technologies for (a) converting coal into liquid and gas, the two fuel forms that pose special resource problems, and (b) extracting substantially more useful energy from uranium than we do with present-day nuclear power technology. The abundance of some other "unconventional" resources of mineral fuels (such as oil shale) serves to reinforce this conclusion on the physical adequacy of domestic resources. So does the possibility of applying solar energy (and various other renewable resources) in certain practical uses.

Future Production Costs

The critical issue shifts then to the matter of future costs in producing useful energy from these abundant resources. In the RFF study, costs for liquids and gases based upon coal and oil shale are estimated at about twice the 1978 prices of petroleum and natural gas. For electricity from coal and uranium, we think that future production costs may avoid a rising trend (after adjusting for general price inflation) except for continued pressures arising for environmental, health, and safety reasons; but those pressures are impossible to predict.

A major influence on long-term costs will result from the fact that in many of the new technologies natural resource costs should become relatively less important while the costs of plant and equipment become relatively more important. Increasingly, then, fuels will shed the cost characteristics of an extractive commodity and take on those of a manufactured product. This would be particularly true of nuclear energy produced in breeder reactors, for which the costs of raw uranium could become a vanishingly small element per unit of energy production.

In processes based on coal and oil shale the raw fuel component would continue to be significant; but even so, the cost of useful oil and gas produced from them will be determined less by rising extractive costs over time than would have been the case for the petroleum and natural gas replaced. For example, even if the cost of mining coal doubled, this factor would mean only a one-third increase in the cost of producing synthetic oil or gas from coal. Costs for the synthetic and natural sources of oil and gas should tend to converge over time, and cost increases from that point on should be more gradual for the synthetics than they would have been for the natural sources at equivalent output levels.

After technologies of this type mature, their long-run cost profiles are quite likely to take the general form of a plateau—now sloping upward with added environmental or raw material costs and now downward with technological advances—rather than a series of sharp steps upward. Thus, their cost behavior should depart markedly from the common expectation of sharply rising costs for industries based upon depletable resources. This is a crucially-significant element in the long-term outlook, and it would, of course, be rendered even more favorable by commercial success with solar energy and other renewable resources.

Outlook for Consumption

I want to turn now to the outlook for energy consumption. Compared to a 1976 level of about 75 quads, the RFF study estimates consumption at about 115 quads for the year 2000 for an economy that grows at a rate consistent with current expectations. This implies an annual average energy growth rate of 1.8 percent compared with an estimated GNP growth of 3.2 percent per year. In other words, we find that the total economy can grow at a substantially faster rate than the use of energy. (This, incidentally, would not be a new experience in the United States;

during the long period between the end of the First World War and mid-century, energy grew generally at a slower rate than the economy, the ratio of energy to GNP falling by about one-third over the entire period—and this during a time of abundant energy supplies at low prices, when, on economic reasoning, energy intensity should have increased. Why this happened, contrary to what simple economic models would suggest, poses an interesting question for combined economic and engineering analysis.)

Our estimated level of energy consumption for the year 2000 is considerably less than such an estimate made prior to the energy upheavals of the past few years would have been. Without taking into account the new realities on energy prices and conservation that have surfaced in recent years, estimated energy consumption in the year 2000—within the same economy-wide assumptions—would have been about 135 quads, instead of 115. The estimated annual amount of energy saved—20 quads—is not negligible. It is equivalent to all of the natural gas consumed in the United States in 1976, or about 1½ times the coal consumed in that year.

Let me illustrate the types of calculations that entered into our reduced consumption estimate with two important examples—private automobiles and residential heating.

- With the number of automobiles expected to increase substantially between 1976 and the year 2000, we nevertheless estimate that automotive transportation will consume *less* fuel in the year 2000 than it did in 1976—dropping by about 30 percent, from slightly over 10 quads to about 7 quads. Our estimates are for about 150 million vehicles by the end of the century (up from about 110 million now).

  There are various indications that we are approaching a substantially slower growth in the total number of vehicle miles traveled, and the projected rise in that crucial figure is more than offset by estimated improvements in gasoline mileage. We estimate the fuel efficiency of the total fleet of automobiles on the road in the year 2000 at 27½ miles per gallon, the average mandated for cars to be produced in the 1985 model year. (The 1976 fleet average was about 14 miles per gallon.) This doubling in fuel efficiency would, incidentally, just about offset the doubling in liquid fuel prices we estimate for synthetic technologies based on coal or shale, and make it that much easier to absorb the cost rise.

- Housing units are also expected to increase substantially between 1976 and the year 2000, from about 70 million to roughly 115 million. Nevertheless, the fuel consumed in residential space heating is estimated to *decline* from just over 8½ quads to about 8. Our housing estimates imply more commodious occupancy levels of 2.3 persons per dwelling unit in the year 2000, compared to 3 per unit in 1976, and more floor space per dwelling in new houses, in line with recent construction experience.

  Even with 45 million more dwelling units and improved space standards, however, we estimate that thermal improvements in the housing shell (including the use of passive solar techniques in new construction) and increased efficiency in heating systems might actually diminish the total fuel consumed. These estimated increases in efficiency could be achieved through improvements which, on balance, result in cost savings to consumers, again making energy cost increases easier to absorb.

Even though the conservation practices built into our estimates yield substantial energy savings, we anticipate a 40-quad increase in the level of energy consumption in the year 2000 compared to 1976. This change implies cumulative consumption over the intervening quarter century of about 2,400 quads (an amount that is equivalent to considerably more than *all* the estimated remaining recoverable domestic resources of conventional oil and natural gas).

Energy consumption probably could be reduced even further by techniques that would enforce higher fuel efficiency standards through various regulatory approaches. In our estimates, we avoided pressing fuel efficiency as the decisive criterion for determining government policy in the energy consumption sphere, because we wanted to avoid the negative repercussions on economic growth and productive efficiency that could result from a single-minded pursuit of fuel efficiency targets to the neglect of other important objectives.

The Energy/Productivity Connection

Of particular concern is the relationship—not yet sufficiently well understood by economists—between the use of energy and the productivity of labor and capital employed in productive processes. The productivity issue is vitally important today as the American economy finds itself straining—and failing—to meet a variety of national objectives. The annual dividend yielded by economic growth has been dwindling in recent years, and the decline of productivity growth is a root cause.

In evaluating the connection between energy use and the efficiency of productive operations, it helps to divide energy consumption into two broad categories: (a) energy as a consumer good, as in the heating of homes and the fueling of private automobiles; and (b) energy as an input into the productive sectors of the economy, such as industry, agriculture, and services. Although these two broad categories overlap (private automobiles do, after all, transport people to their jobs and to stores and other places of business), the general distinction is useful.

Where energy is a consumer good in itself, it is difficult to argue that reducing its use (particularly where guided by considerations of cost-effectiveness) would jeopardize the efficiency of productive operations. Energy used for personal purposes is primarily one of the proceeds of the productive economy, rather than a factor in national output. This does not mean, however, that consumer use of energy can be dealt with cavalierly. Questions of welfare and equity are involved, particularly for those who are just beginning to enjoy comforts that were previously beyond their means, as is the link between personal energy use and the activities of the productive sectors of the economy.

Energy as an input into productive operations is an essentially different matter. There the quantity and form of energy used have in the past decidedly affected the efficiency with which labor and capital have been combined to produce output. Even when it is cost-effective to substitute other productive factors for energy, the nation may wind up with reduced productivity as a result. If energy costs rise, production may be optimized by using more labor and less energy; but the total output of goods and services may well be lower than it would have been if energy were used more freely at a lower price. Reducing energy use even further—by imposing taxes on its use or through other policy instruments that might be desirable for many reasons—could compound the problem of losses in productivity. Fuel efficiency should therefore be pursued, but with full regard to its implications for the overall performance of the national economy.

Policies for Production and Use

I turn now from the "yes" part of this so-far generally optimistic response to a consideration of the second part of the proposition stated as the thesis to be examined in this conference, the part that states

". . . provided that proper policies and implementations are arranged to ensure a plentiful supply of energy at the assumed higher prices."

First, I want to offer what I regard as an important addition to the statement just quoted: proper policies are needed not only to ensure a plentiful supply of energy, but also to assure that energy is not used wastefully. And to ensure this latter objective the most important policy of all is to get energy prices right. I am sure that this point is by now well understood. But also well understood are the great difficulties connected with the implementation of such a policy—its implications for inflation and for the inevitable concerns about "fairness" raised by increases in energy prices. This is as difficult an area as any in the entire sphere of energy policy, but movement toward implementation has at least been started.

I do not yet see a similar forward movement in connection with the critical elements of a coherent policy that will permit a timely transition to technologies that would make energy supplies abundantly available on favorable cost conditions over the long run. Some things can be achieved by "getting prices right" and letting decisions be guided by the market place. This should be helpful in encouraging the expansion of output of conventional sources of petroleum and natural gas.

However, the supply technologies that offer the greatest hope for the long run cannot, for the most part, meet the test of the market place without a period of sustained government assistance and, the decisions to provide the necessary support will not be easy. Aside from the budgetary problems that might be posed, there is the fundamental difficulty that some of the most attractive technologies from a cost and supply standpoint have aroused serious opposition on health, safety, and environmental grounds, while most of the technologies that appear to be benign do not look promising from a cost standpoint.

This dilemma is perhaps the chief factor standing in the way of decisions on a long-term supply strategy. But even so, there are hopeful aspects. In significant ways, new supply technologies based upon coal and nuclear resources can be less damaging to health, safety, and the environment than those they would replace. Nuclear power, in its routine operations, would cause fewer fatalities and illnesses, and less damage to land, than the chain of supply required to get electricity from coal. New technologies for using coal directly and for converting coal to liquids and gases should be less polluting than the old coal-burning processes they would replace. Some solar technologies appear to be less damaging than either coal or nuclear and this may justify a subsidy, representing their social benefits, which would help to offset their monetary cost disadvantages.

That is the hopeful side. But there are also persistent concerns about the catastrophic threats that nuclear power poses in the form of possible major reactor accidents, and through the possible international spread of weapons capabilities because of the easier availability of plutonium, particularly if breeder reactors and fuel reprocessing are used. There are also concerns about the long-term implications for climate of a build-up of carbon dioxide in the atmosphere because of the expanding use of fossil fuels. Even if the probabilities are slight that some of these events will happen, the effect of the "worst case" could be devastating. On the other hand, there are institutional and technical approaches that might reduce some of these risks to acceptable levels and these approaches could be pursued vigorously.

America's Energy Goals

After evaluating the major factors bearing on supply, demand, environment, health, and safety, the RFF study concluded that it should be possible to achieve a consensus in support of a long-term national energy strategy that explicitly combines three goals: domestic supply adequacy consistent with requirements for economic growth; conservation in energy consumption; and protection of health, safety, and the natural environment. The hopeful message of the study's findings is that feasible, identifiable supply and conservation technologies might enable us to achieve major elements of these *combined* requirements over the long run. Not only are such objectives not mutually inconsistent; they may indeed dovetail in significant respects. Still, practical outcomes would require numerous interrelated policy initiatives; and some of them would need to begin soon because of the very long lead times involved in producing results, particularly where new supply technologies are concerned.

It is becoming increasingly clear that we are running out of time for devising and implementing policies to meet long-term energy needs. Regardless of the nation's wealth in energy resources, we stand to suffer serious disruptions unless the necessary decisions are made—and soon. Agreement on broad national goals is essential, not only to provide a starting point for launching timely initiatives to meet long-term needs, but also to aid in coping with short-run problems. Realistic long-run objectives, which identify the nation's energy capabilities and the means for achieving chosen ends, and which place the short-, medium-, and long-term in proper perspective, could serve to dispel the pervasive feeling that this country can do no more than react passively to developments elsewhere in the world—actions and situations over which it has no control.

Discussion

From the floor: Relative to economic tradeoffs in the present and future costs of power, it might be noted that the geysers in northern California have been producing power since 1960, and the total investment to date for the approximate 600 megawatts of power that have been produced is somewhere between three and four hundred million dollars. Is this not a very acceptable cost of energy when we relate it to current nuclear power plant and fossil fuel costs? Why is geothermal energy, in view of this long record, being so underplayed in our future energy plans?

Schurr: I assume that geothermal is being used in California because it is economic and has made itself a part of the ongoing competitive market on cost considerations alone. Wherever those conditions could be duplicated, there would seem to be ample justification for proceeding with geothermal. The question is, how widespread *are* the locations that repeat the kinds of favorable circumstances that exist here? If they were sufficiently widespread so that one might regard geothermal as an energy resource like coal or nuclear that can be made available anywhere, then that obviously would be a major element to be considered in future planning. To the extent that geothermal is limited by site specificity, its implications for the nation as a whole are that much smaller. I should think considerations of that kind must play a part in evaluating the comparative importance of geothermal in relation to other energy sources that can either be transported like coal and uranium or are available ubiquitously like sunlight.

Charles Hitch (from the floor): It seems to me that there is not necessarily a difference between the price projections that Kenneth Davis used—an average 50 percent increase in energy prices to the consumer [using our domestic resources] and your doubling of the prices of *primary* energy. They are quite consistent with each other.

Schurr: You are probably right. As a matter of fact, Ken Davis pointed out to me that he was talking about delivered prices whereas my estimates were tied to the cost of production. Taking distribution, transportation, and so on into account, the two projections might work out to be quite consistent.

Henry Linden (from the floor): Would you explore further your comment on the relationship between productivity and the inputs of labor, capital, and energy? United States industry since 1970 has increased its output by 30 percent in relation to energy input. Is there a rational system of adjusting the employment of energy if it is economically sound? Are we over-conserving in United States industry? Is this in any way related to the productivity decline?

Schurr: You have raised a number of serious questions to which the answers are not available. It has been argued by some economists that there is a connection between the recent dismal behavior of productivity and recent increases in energy prices that have led to a shifting in the way in which factors of production are combined. There is, however, no consensus on this point that I am aware of; I think it is still under discussion and debate.

My own observations were based on the longer historical record where it can be shown that the conditions of energy supply—ample availability and low prices—played a significant part in facilitating certain major developments in productive operations in manufacturing, agriculture, and other sectors, which were then instrumental in raising the level of productivity in those sectors and thereby in the total American economy. One needs to keep that in mind in considering what the implications are of going to a regime of higher energy prices or restrained availability. The specific elements in this past energy-productivity connection can be quite readily documented.

With this in mind, and looking to the future, I think it is not sufficient to do what a lot of economists do, which is to say there is obviously some elasticity of substitution between energy and other factors of production, and that the country will therefore get by quite satisfactorily with a smaller amount of energy at higher prices because it will then pay to substitute other factors for energy. It may pay to substitute other factors for energy—that is a perfectly reasonable observation to make—but that does not complete the loop of significant effects; to complete the loop it is necessary to understand what effects changes in the combination of factors of production will have on the efficiency of productive operations. Taking the historical record as my guide, I find myself greatly concerned about this; I cannot accept with as much equanimity as many other economists do the notion that the problem is solved if the elasticities are known. Of course, the elasticities are not known; the data base is necessarily inadequate because it refers to a period in which prices were low and falling, not high and rising. But even if the precise figure for the elasticity of substitution were known, that would still not give economists the whole answer, because the important feedback on productivity would need to be taken into account.

To my mind, concern about the potential impact of energy use on productivity is significant enough to argue, at least, against doing those things that cannot be justified on grounds of cost effectiveness. To go beyond cost-justified substitution might really be flirting with danger. But even to go no further than cost-justified substitution, which is where the economy would lead under ordinary market operations, does not offer assurance that there will not be detrimental effects on productivity that are not measured in existing econometric models.

From the floor: You concluded your talk by saying we must come to an agreement on our goals. It is obvious that nationally we are tearing ourselves apart in opposite directions, and that some educating should be done. The public should be informed as to the facts, so the decision-making process can benefit. Who, in your opinion, should be doing this educating?

Schurr: On the assumption that responsible scholarship uncovers the underlying factual base, it would be ideal if one could get the media of communications to present such information in a form that is less concerned with the dramatic and catastrophic, and more concerned with trying to understand the choices that are available to us and how they compare. The public needs to know what is gained, what is given up, if one alternative is chosen over another. It also needs to understand how different time phases tie together—the short-run, the intermediate-run, and the long-run. Everything cannot be done at once.

There is also a leadership role that the President and others in public life could play to set forth and emphasize underlying factors that must be considered. People in political life, however, are so much concerned with the day-to-day that it is often difficult for them to spend time trying to place problems in a broader and deeper perspective.

There is also a role for academic and research institutions in providing the underlying information, although they cannot be expected to disseminate it; that is not their special strength. Certainly, however, they might be expected to help uncover and compare the elements of choice available to us as a nation.

William W. Hogan

Economic Growth and Higher Priced Energy

The relative cost of energy changed dramatically in 1973-74, creating a new era of gloomy expectations about a future of higher priced energy. The age of limits was said to be upon us, with the energy crisis labelled as part of a pattern of forces that will permanently constrain economic growth.

This was evidently true in the short run—the severity of the recession of 1974-75 can be attributed to the sharp increase in oil prices and the policy response to constrain inflation. But this immediate dampening of economic growth need not be matched by a similarly severe reduction over the long run.

The analysis of limits and the most pessimistic interpretations of the effects of higher energy prices overlook the role of energy in the economy and the substantial opportunities to adjust the utilization of energy.

The amount of energy required for any level of economic activity can be varied, then the increase in energy prices may lead to shifts in the use of energy and other resources, with a significant but not catastrophic reduction in overall productivity.

The purpose of this paper is to summarize an analysis of evidence on the economic role of energy. First we examine the aggregate contribution of energy to the economy and conclude that energy, although

William W. Hogan *is professor of political economy at the John F. Kennedy School of Government, Harvard University, holding the IBM Chair in Technology and Society. He founded and is Executive Director of the Engineering Modeling Forum of Stanford University. In 1975-76 he was Deputy Assistant Administrator for Data and Analysis of the Federal Energy Administration, and directed the Office of Quantitative Methods.*

pervasive and important, plays a relatively small role. Then we consider the possibility of adjusting our utilization of energy, i.e., we estimate the potential for energy conservation, and determine that the weight of evidence supports the view that there is a great long-run potential to improve the efficiency of energy use.

It follows, therefore, that a high rate of economic growth is possible in a world of higher priced energy, once the adjustments to the new price regime have been made. And if the prices stabilize, with abundant supplies at the high prices, the economic component of the energy crisis should recede into the background of active policy concerns.

Energy Use

Until recently, energy policy makers expected the aggregate demand for primary energy, measured in terms of the equivalent heat content of oil flowing from the well or coal extracted from the mine, to grow steadily. Prior to 1970, there were abundant supplies of cheap energy. Markets for new energy-using technologies, such as air-conditioning, were expanding rapidly, and the growth in energy demand closely paralleled the growth in the gross national product. Between 1950 and 1970 the United States economy grew at 3.2 percent per year; the demand for energy grew at 3.4 percent per year. This close linkage between energy and the economy led to a widespread presumption that growth in energy consumption was essential for the future growth and health of our economy.

Of course this period was also a time of great change in the energy system. The relative importance of coal decreased sharply, with petroleum and natural gas taking its place—the fraction of total energy consumption coming from coal dropped by half over this period, while the proportion coming from natural gas increased by nearly the same amount. The form of energy use changed as well, with the energy going to electricity generation increasing from less than one-sixth to nearly one-fourth of all the primary energy used. These changes were of great importance to the energy industry, but passed unnoticed or were taken for granted by the average energy consumer.

By 1970, however, the energy system began to change its complexion in ways that affected our lives and our perceptions of the future. The regulation of natural gas prices became an increasingly controversial issue, and the first curtailments of natural gas consumption had begun in order to allocate limited supplies in the face of a growing demand. The domestic production of crude oil began to decline, with natural gas production soon to follow. The environmental problems associated with burning of fossil fuels had moved to the forefront of national concern; for example, the regulation of power-plant emissions had become a major element of national policy. Finally, the events of 1973, with the oil embargo and the abrupt increase in the price of oil, created a new concern with the energy-economy and a fear that the future would be one of energy scarcity unlike the past of energy abundance.

This new awareness of the importance of energy has led to a reevaluation of both the role of energy in our economy and forecasts of the likely future demand for energy.

At one extreme, there is the view that energy consumption is closely linked to the level of economic activity and any expansion of the economy requires a proportional increase in energy supply. From this perspective, which seems supported by the United States data from the years prior to 1970, economic growth determines energy demand, which must be treated as a requirement to be met through the sacrifice of great resources, the compromise of conflicting environmental goals, or both. At the opposite extreme, there is the view that the increased use of energy is only one of many ways to accomplish our real goals, and energy demand can be reduced without great sacrifices in welfare, preserving our limited energy supplies for the future and improving the environment today.

The evaluation of the potential flexibility of energy use will have a great influence on the types of policies recommended for our energy future. Those who see energy demand as inflexible will favor dramatic programs to expand energy supply, even to the point of subsidizing its production and use. Those who see more flexibility in the way energy is used will choose the energy conservation policies that could lead to sharp changes in our energy-use patterns.

Our focus is on the long run, when many adjustments in economic structure can occur. In the long run, higher energy prices will stimulate energy conservation, reducing the use of energy and moderating the economic impact of energy scarcity. It is this change in energy use patterns that is the key to understanding the link between energy prices and economic growth. Despite their importance in the short run, the long-run analysis abstracts from the problems of inflation and unemployment.

Assuming that the economy is managed well, therefore, we concentrate on a few questions: how important is energy in the overall economy; how much will higher priced energy induce energy conservation; how will the reduced use of energy affect the potential output of the economy?

The Elephant and the Rabbit?

The long-run link between energy and the economy has been studied intensively in recent years, and many detailed models have been built to provide insight into the problem. The issue is scarcity, with higher energy prices and the impact on the full employment potential of the economy.

None of the models implies that energy is not needed for the economy. If energy prices remain stable, then an increase in economic activity should produce an increase in the demand for energy. This would explain the history of energy and economic growth when the prices were low.

Of course, the future growth of energy demand may be less than the historical growth because of lower projections for population increase or a trend toward a disproportionately higher growth of the less energy-intensive sectors of the economy. The difficult question is, can energy demand growth be dampened further by higher energy prices without proportional reductions in economic activity?

For simplicity in our analysis, we represent the economy in terms of just two inputs—energy and all other items. The energy input is measured as closely as possible to the point of production, i.e., oil at the wellhead. This does not include the capital and labor needed to refine and distribute the energy. Presumably, if energy consumption is reduced, in the long run this capital and labor could be allocated to other uses; valuing energy at the point of production avoids the double counting.

Note in Figure 1 that energy is only a small component of the total U.S. economy. As of 1970, the value of primary energy inputs did not exceed 4 percent of the GNP. The analogy of an elephant-rabbit stew illustrates the implications of this low value share. If the recipe for such a stew calls for just one rabbit (the energy sector) and one elephant (the rest of the economy), won't it still taste very much like elephant stew? If energy prices had not risen after 1970, it is likely that energy demands would have grown at about the same rate as the GNP. The 4 percent ratio would then continue into the future.

But what would be the effect if energy costs were to double and yet there was sufficient time given to the economy to adapt? One estimate of the impact may be obtained by assuming a constant recipe. Suppose the

Fig. 1. GNP and Energy

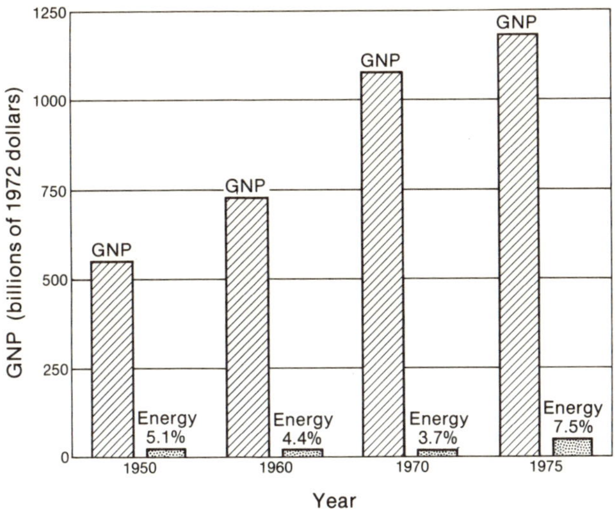

Source: "Energy and the Economy," Vol. 1, Energy Modeling Forum, Stanford University, 1977.

rabbit is paid for with part of the stew. Then an additional 4 percent of the stew (GNP) must be allocated to cover the doubling in the cost of the rabbit (energy). In fact, other recipes are available that call for less rabbit (energy conservation) and, therefore, lead to lower costs. Under these assumptions, the first doubling of energy costs would produce, at most, a 4 percent loss in GNP.

With a more complicated argument, it can be shown that a small decrease in energy supply leads to a decrease in economic output proportional to the value share of energy in the economy. At a 4 percent value share, a 1 percent reduction in energy input would produce a 0.04 percent drop in total output. By this argument, a small percentage change in energy availability produces a considerably smaller percentage impact upon the economy as a whole.

This simple analysis provides some insight, but it suffers from a major defect in failing to represent accurately the flexibility of energy utilization in the economy. The processes for utilizing energy are not fixed immutably. Within limits, insulation, efficiency improvements, and changes in the mix of input factors can alter the energy requirements for a fixed level of output. Such substitution possibilities can modify the economic impact of changes in the energy system.

Flexibility through Substitution

The specific processes for energy substitution may be varied and intricate. Therefore, even if it is generally agreed that some substitution is feasible, it may not always be possible to identify all the specific technological options available. The detail may be approached gradually by expanding our first simple analogy based on the value share of energy.

We explicitly assume now that substitution is possible between energy and nonenergy inputs to the economy. As energy prices increase, the cost-conscious energy user will reduce the use of energy and substitute other resources, such as capital and labor. The end result, after all the many adjustments have been made in the economy, can be summarized by the proportional reduction in energy use, for a given level of output, in response to a change in relative energy prices.

Elasticity of Substitution

To a first approximation, this measure of flexibility can be represented in economists' terms as the elasticity of substitution. Ignoring the feedback to the economy or other inputs, this parameter is the same as the elasticity of energy demand. The elasticity of energy demand measures the percentage response of the demand to a given percentage change in energy prices. Hence, if the elasticity of demand is 0.3, a 10 percent increase in energy prices produces a 3 percent decrease in energy demand, a decrease achieved by substituting capital and labor for energy.

Of course, it is easier to define the aggregate elasticity than it is to identify the many price-induced changes in individual energy use patterns or measure the elasticity that these changes imply. (The measurement problem is addressed below in the examination of the historical evidence for different economic sectors to provide a range of estimates of the elasticity of demand.) But, if the measurement problem can be overcome, the aggregate elasticity can be used to examine the impacts of energy scarcity.

This concept, the elasticity of substitution, provides a convenient index for summarizing the aggregate behavior of the detailed energy-economic models. If we assume that inputs of other factors such as capital and labor are held constant, then the elasticity of substitution virtually determines the feedback effect of the energy sector on the rest of the economy.

The implications of alternate elasticity estimates (denoted as σ) are shown in Figure 2, which is drawn from a comparative study of energy-economic models. This depicts the GNP in the year 2010 (long enough in the future to ignore transition effects) as a function of energy input, holding other inputs constant. It is assumed that a Btu tax is imposed gradually to reduce energy consumption, with the revenue rebated to consumers. Such a tax might be levied, for example, to mitigate environmental impacts or lessen important vulnerability.

Fig. 2. Economic Impacts of Energy Reductions in the Year 2010 for Various Elasticities of Substitution (σ)

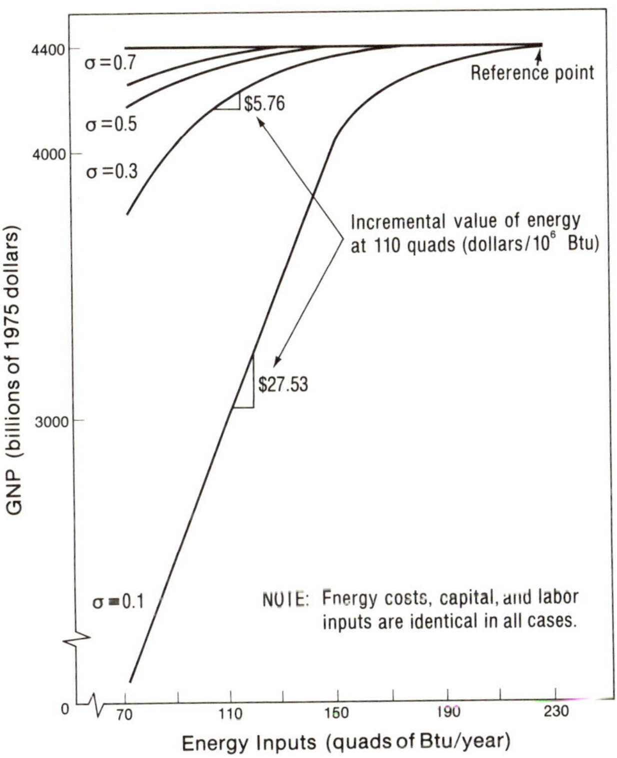

Source: "Energy and the Economy," Vol. 1, Energy Modeling Forum, Stanford University, 1977.

The analysis is based on the hypothetical reference forecast, but the qualitative conclusions are not sensitive to the reference assumptions. A small change in energy availability has almost no effect on GNP. The loss in output is exactly balanced by the savings from the reduced payment for energy. This is what the price represents in a competitive market: the value of the product at the margin. This value will change as the quantity of energy input changes, but the economic output does not decrease in proportion to the decrease in energy input. Substitution of other input factors, such as capital and labor, compensates for the reduction in energy supply.

The importance of the long-run elasticity of substitution is startling in the context of this analysis. A 50 percent reduction in energy availability produces a 28 percent reduction in GNP if the elasticity is as low as 0.1 but only a 1 percent reduction in GNP if the elasticity is as high as 0.7. Seemingly small changes in the substitution potential produce major changes in economic impact.

Even the smaller GNP reductions have a large value, however. If the economy is growing at 3 percent in real terms and we discount future consumption at 6 percent, then a 1 percent reduction in annual GNP corresponds to a present value of nearly half a trillion dollars. This is only 1 percent of the present value of future output, but it would justify a substantial research investment aimed at developing low cost technologies which can expand energy supply or improve the efficiency of energy utilization.

The estimates in Figure 2 of the impact of energy reductions are based on a simplified, partial analysis of the economy. This shows the potential of substitution between economic inputs to absorb energy input reductions with less than proportionate reductions in economic activity. However, changes in energy input have a further dimension of feedback to the economy. This additional dimension centers on the pattern of capital investment over time. Reductions in energy input lead to changes in the rate-of-return on capital as well as reductions in the level of total output. Investment, savings, and capital use are altered as a consequence. Over time, these effects may accumulate into significant changes in the stock of capital and, therefore, in the productive capacity of the economy.

Changes in Capital Input

It is difficult to analyze these complex interactions; sophisticated models are required for this task. As a good approximation, however, we can extend the partial analysis to illustrate the magnitude of the economic effects of changes in capital input.

For this purpose, expand the beginning framework to include three economic inputs—energy, capital, and labor. Now, instead of holding both capital and labor inputs constant as energy input changes, let capital adjust to maintain a constant rate of return. Since total labor input remains unchanged, the returns to labor are reduced. The impact of this new assumption, for the case of an elasticity of substitution of 0.3, more than doubles the feedback effect. For a 50 percent reduction in energy input, there is a 4 percent reduction in GNP when capital is held constant, but the reduction is 11 percent GNP when energy is reduced and capital changes to maintain a constant rate-of-return. In this case, the capital change effect exceeds the direct effect of the energy reduction.

Thus, both substitution and capital adjustment processes are important in considering the feedback effects of energy on the economy. When both these processes are taken into account, however, the conclusion remains: while energy reductions do have a substantial economic impact, the GNP reduction is proportionally smaller than the reduction in energy use and is sensitive to the elasticity of demand. It is important to examine the data to estimate this critical aggregate elasticity of demand.

Estimating Substitution Potential

The profit-maximizing firm and the cost-conscious energy consumer will consider the prices of all inputs when making decisions about the mix of goods and services to be utilized. As the price of energy increases, we assume that energy users exploit the opportunities to substitute more energy-efficient machinery, to redesign production processes, to insulate homes, to install heat pumps, to switch to buses and subways—even to lower thermostat settings and observe speed limits; the energy user makes these and many other substitutions in order to reduce the cost of meeting the more fundamental objective of producing a product or achieving an acceptable standard of living.

The degree to which these adjustments will be made depends on many factors. The technical limits are constraints, but the actual substitutions may be motivated as much by convenience and personal taste as by the engineering characteristics of specific energy uses: the existence of hand-operated can-openers does not imply that cooks will or should abandon the electric counterpart. The measurement of the energy substitution potential, therefore, must be based on the observed behavior of energy users confronted with different prices and the opportunity to make consumption choices. The result, if not the process, can be summarized by the proportional reduction in energy use for a fixed level of activity, in response to a given percentage increase in the energy price, i.e., by the elasticity of energy demand.

The aggregate elasticity of energy demand will be useful in summarizing the substitution potential, but it is not useful for actually estimating the degree of potential substitution. Energy is not a single commodity and the full richness of the energy system must be considered when estimating the elasticity of energy demand. And only the most general linkage can be established between the disaggregated estimates of the many adjustments, made by different users, confronted with different prices, for different energy products, and the abstraction of the aggregate elasticity of total energy demand for all uses.

Difficulties of Measurement

The links between the aggregate elasticity, as used so far in this paper for analytic convenience, and the elasticities for individual sectors or fuels are obscured by the diversity of prices that must be considered. It is the price to the user that is the relevant price in determining substitution among the many energy and non-energy inputs for production or consumption. This is a weakness of analyses measuring the price of energy in terms of the price of primary energy. No consumer actually purchases energy at the price of production, and the prices to consumers do not vary in constant proportion with the price of primary energy. Hence the same absolute price change and the same demand response will lead to different estimates of the price elasticity depending on the point of measurement selected.

For example, in 1972 the price of oil at the wellhead accounted for slightly more than one-fifth of the cost of gasoline at the pump. Hence, assuming a straight pass-through of increased crude oil costs and no change in refinery operations, the elasticity of demand measured at the pump could be five times the elasticity measured at the wellhead.

Despite, or perhaps because of, these difficulties of measurement and interpretation, there have been many different attempts to examine the data and estimate the elasticities of energy demand. Many factors must be controlled (e.g., weather, population, economic activity) to isolate the effects of price alone, and the short-run effects must be separated from the long-run adjustments that may take many years to complete. In addition, the data are not perfect and there is no guarantee that the changes of the past, over one range of prices, will be indicative of the changes in the future, over a very different range of prices.

The results are not without controversy, therefore, but some patterns do emerge. One survey of a number of different studies compared the long-run elasticities of demand for different consuming sectors, as shown in Table 1. These figures indicate a substantial elasticity of energy demand, i.e., a significant flexibility to substitute for energy use.

Table 1. Estimates of Long-Run Elasticity of Energy Demand *(Delivered to the Consumer)*

Sectors	Range of Studies	Most-Likely Range
Residential	**.28 to 1.10**	**.7 to 1.1**
Industrial	**.49 to .90**	**.8 to .9**
Transportation (Motor Gasoline)	**.22 to 1.3**	**.7 to 1.0**

Source: R. Pindyck, "The Characteristics of the Demand for Energy," in J. Sawhill, ed., *Energy Conservation and Public Policy,* Prentice-Hall, New York, 1979. The survey by Pindyck covered several detailed models and econometric studies, with qualitative judgments used to narrow to the most-likely range.

These aggregate elasticities are roughly consistent with the potential reductions in energy utilization found in detailed technical analyses of individual conservation opportunities. The implication is that a doubling in the real cost of energy delivered to the consumer could produce up to a 50 percent reduction in the intensity of energy utilization. Of course, this would require more than a doubling of the cost of primary energy, but this represents a substantial flexibility in the long-run demand for energy, consistent with a σ of .4 to .5 in the aggregate analysis above. And these aggregate figures disguise even greater adjustments in the demands for individual fuels; the increase in the price of any individual fuel will cause a shift away from that fuel, in addition to a reduction in the total energy demand.

These are long-run effects, however; the timing of the adjustments may produce very different effects in the short run.

Energy Demand Adjustments

After many years of declining real prices and relatively stable demand growth in the energy system, in 1973 the sharp increase of oil prices marked the beginning of what may be a long period of change and adjustment in the level and composition of energy demand. Reinforced by the recognition of environmental problems and complicated by a high rate of general inflation, these adjustments were still very much underway more than five years later. Any investigation of future energy demand must consider both the continuing adaptation to past changes, e.g., in the price of energy, as well as the effects of future rates of increase of energy prices and economic activity.

Our exploration of the possible energy-demand adjustments will be improved by the refinement of our measure of aggregate energy consumption; we need a yardstick that can be related easily to the data and yet interpreted as an index of the final consumption level of the energy consumer. The difficulties inherent in the development of such an index are apparent: it must provide a consistent aggregation of the diverse energy products, from the coal consumed in industry, at prices as low as $0.80 per million Btu's, to the electricity consumed in the home, at prices as high as $10.00 per million Btu's delivered. It is not possible to specify a perfect index, but it is possible to construct a good approximation through a careful accounting for the different prices paid for different energy products.

The aggregation employed here is the same as the *net-energy* consumed: that is, the equivalent-heat content, delivered to the consumer, net of the energy losses in the generation, transmission and distribution of electricity. The prices are those paid by consumers, as opposed to wellhead or primary-energy prices, aggregated across sectors, weighted according to the value shares of expenditures.

One advantage of measurement at this point is that it is the best approximation to the point where energy choices are actually made by the consumer. A disadvantage is that the resulting energy demand total is not the same as the total primary-energy demand. The energy losses in electricity generation must be added to the net demand in order to obtain the equivalent primary-energy demand.

Net-Energy Demand/GDP Ratio

If the scale of economic activity is assumed not to affect the composition of energy demand, then the major uncertainty associated with economic forecasts can be avoided in the analysis by concentrating on the ratio between the delivered net-energy demand and the Gross Domestic Product (GDP).

The data for the United States and selected industrial nations are summarized in Figure 3. The net-energy-GDP ratio for the United States is higher than that for most other countries but has dropped steadily,

Fig. 3. Net-Energy-GDP Ratio: Selected Industrial Countries

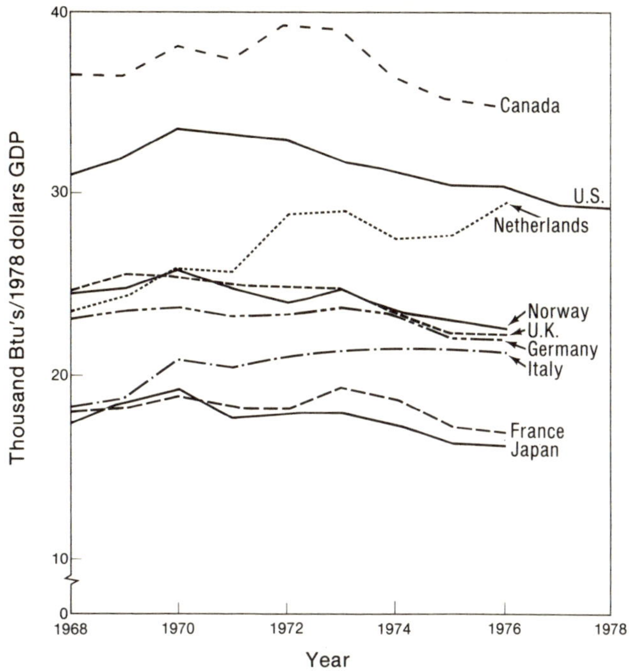

Source: Quantity data are from OECD *Energy Balances: 1974-1976*. For details, see W. Hogan, "Dimensions of Energy Demand," in *Selected Studies on Energy: Background Papers for "Energy: The Next Twenty Years,"* Resources for the Future, Washington, D.C., 1979.

from 33.4 thousand Btu's per 1978 dollar of output in 1972 to 29.1 thousand Btu's per 1978 dollar of output in 1978. The data from the other countries do not show quite so regular a pattern, but there is a general, if slight, reduction in the overall use of energy.

Part of the explanation for this improvement in energy efficiency can be found in the trends in energy prices. The aggregate energy prices for the consumers in each of these countries are displayed in Figure 4. There has been a significant real increase, since 1972, in the price of energy, although the increase is much less than many would expect given the dramatic changes that have been experienced for the prices of some important energy components; e.g., for oil in the United States which increased by 120 percent between 1972 and 1978. The difference can be explained by noting that oil is only one component of aggregate

Fig. 4. Price of Net-Energy Delivered: Selected Industrial Countries

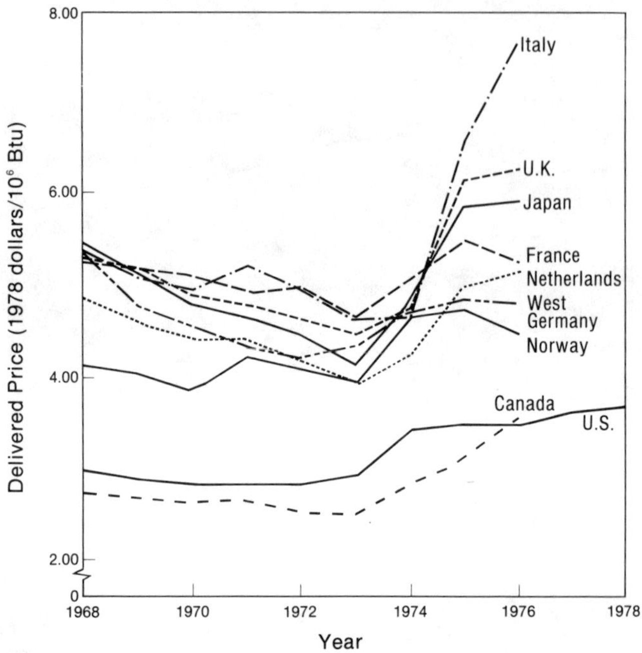

Source: See W. Hogan, "Dimensions of Energy Demand," in *Selected Studies on Energy: Background Papers for "Energy: The Next Twenty Years,"* Resources for the Future, Washington, D.C., 1979.

energy input; many products are still sold under long-term contracts at low prices. And the costs of primary energy forms, e.g., oil, coal, and gas, are only a part of the price of delivered net-energy. Hence, the nearly threefold multiplication of imported-oil prices since 1972 has led to only a 30 percent increase in the delivered price of all energy in the United States.

If we assume that a change in energy efficiency takes place only with the introduction of new energy-using equipment (e.g., only by changing the efficiency of new cars, not by driving less), then the relationship over time between delivered-net-energy prices and the net-energy-GDP ratio can be established once we specify the long-run price elasticity and the rate of introduction of new equipment. The result is a traditional demand curve, one for each year, with the price responsiveness increasing as we look further into the future.

Consider the examples in Figure 5 based on data for the United States, for demand curves for the year 1978. Four curves are shown, one for each combination of two possible long-run elasticities (e) and two possible rates of introduction of new equipment (d). The price elasticities were selected to be compatible with the range of estimates from Table 1 but to examine both high and low values (e = .7; e = 1.0); recall that these are the elasticities in terms of delivered-net, not primary, energy.

Fig. 5. U.S. Net-Energy Demand Curves for 1978: Selected Elasticities and Adjustment Rates

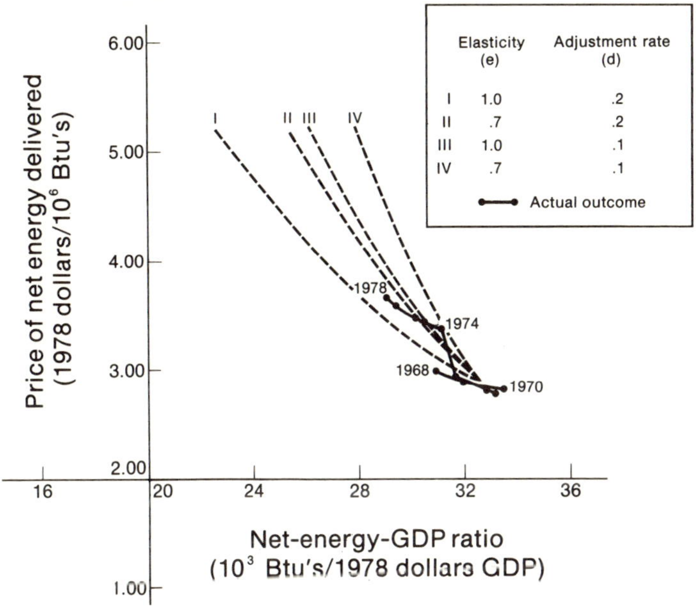

Source: W. Hogan, "Dimensions of Energy Demand," in *Selected Studies on Energy: Background Papers for "Energy: The Next Twenty Years,"* Resources for the Future, Washington, D.C., 1979.

The selection of the rate of introduction of new energy-using equipment is more problematical; two values are considered here (d = .10; d = .20). An introduction rate of 10 percent, i.e., d = .10, implies that in any year 10 percent of all energy-using equipment is newly installed, and is assumed to be at the optimum energy efficiency at the price in that year. An economic growth rate of 3 percent per year and an average life of equipment of 15 years (6 years) would be consistent with a rate of introduction of new equipment of .10 (.20).

The demand curve in Figure 5 for each combination of demand elasticity and introduction rate is based on a forecast from 1969 through 1978, and can be compared with the actual outcomes for the intervening years. For example, in 1978 the net-energy-GDP ratio was 29.1, down from the 1969-1970 average of 32.7; and the predicted value for the high-elasticity-slow-adjustment case (e=1.0; d=.10) is 29.9.

Potential for Change in Energy Use

At least two lessons can be drawn from the data in Figure 5. First, there has been a notable reduction in the net-energy-GDP ratio for the United States, consistent with our expectations given the rise in energy prices. But, second, it is too early to choose confidently, on the basis of these data alone, between the cases of low price-elasticity with a quick adjustment, and a high price-elasticity with a slow adjustment. Of course, even the low price-elasticity is higher than some of the estimates summarized in Table 1.

This short-term evidence, despite its many limitations, supports the view gleaned from Table 1 and studies of the long-run substitutability of energy: there is a substantial potential to change the intensity of energy utilization, and we are going through a period of adjustment to new patterns of energy consumption; further reductions in energy intensity can be expected, even if the final level of demand is in doubt.

If we focus on these two cases, high and low elasticity, we can exploit the same methodology to examine the likely range of energy demand over the next twenty years. By the year 2000, the adjustment to the recent price increases may be nearer completion, but further increases in prices are widely anticipated; hence, we may see an endless process of gradual change, with net-energy-GDP ratios much different in the future than today. For example, using our earlier assumptions, if the price of oil increases to $4.50 per million Btu's, the price of delivered-net-energy would rise to approximately $5.50 per million Btu's. By the year 2000, the corresponding net-energy-GDP ratio could be anywhere from 19 to 22; but in either case, more than 25 percent below the level of 1978. This would be a substantial reduction in the energy intensity of the economy, and would greatly moderate the economic impact of the higher energy prices.

Distribution of Costs and Benefits

The gloomy expectation of increasing energy prices is brightened by the long-run flexibility of our economy. The mitigation of the aggregate economic costs, through substitution and energy conservation, implies that the increasing scarcity of energy need not dominate our future, even though we would like cheaper energy if we could find it.

But energy problems are not likely to vanish soon from the center of policy attention. The proportionally small aggregate economic effects disguise potentially large changes in the distribution of energy costs and benefits, and our political system is very concerned with these transfers of wealth. It may be that the attention to energy and aggregate economic growth is misplaced. Perhaps the real issues center on the distribution of energy costs and benefits—across industries, regions, and individuals.

Across Industries

The distributional effects across industries are understood in principle, and several available energy-economy models can represent separately the impacts of energy prices on different industries.

Energy-intensive industries will be affected most, although less than might be expected. The benefits of substitution and energy conservation are available to dampen the contribution of higher energy prices to the increase of prices for final products and services.

Hence, industry can expect to change its mode of operation, but the long-term readjustment of demand should not produce many dramatic changes in industrial composition. The gasoline guzzling automobile industry will be replaced by the gasoline conserving automobile industry and still be healthy. The energy industry itself will see the most pronounced reduction in demand for its products in response to higher energy prices.

Across Regions

The distribution of energy costs and benefits across regions, which may be the most intractable short-term energy policy problem, is not as well understood.

We know that the economy and the use of energy display a remarkable degree of variation across the continent. There is a range of nine-to-one across regions in the rate of growth of aggregate income and a range of five-to-one across the states in the use of energy per unit of income. The analysis of regional effects is inhibited by the formidable data requirements and by the complex interactions with energy pricing regulations. The energy-economy models do not usually include regional detail, and in the few cases where the attempt is made, there is uncertainty about the supporting data and theory.

Yet the conflicts between the energy-producing and energy-consuming states, the competition for the diminishing supply of price-controlled fuels, and the apolitical migration of energy related pollutants cannot be addressed without recognizing the disparities in the regional availability of energy. An improved regional analysis represents a formidable but important analytical challenge.

Across Income Classes

The distribution of energy costs across income classes is a sensitive political problem, often because energy purchases are believed to be a larger fraction of total expenditures for the poor than for the rich. Of course, the rich will always suffer less than the poor from any broadly based income reductions, but the presumption that energy price increases are highly regressive may be inaccurate.

It is true that direct energy expenditures, the electricity and gasoline we purchase for our homes and cars, are a larger portion of budgets for the poor than for the rich. But energy use is pervasive in the production of other goods and services; therefore, this energy is used by the consumer, indirectly, through the final purchases of goods and services.

When the total purchases of energy, both direct and indirect, are considered, the picture changes substantially. As shown in Figure 6, there is only a slight saturation of total energy purchases for the higher income groups; therefore, an increase in energy prices yields something more like a proportional decrease in income for all groups than a decrease directed primarily at the poor.

The income distribution effects of more selective changes in energy prices are not so clear; e.g., a tax on gasoline might be regressive, but deregulation of natural gas prices might affect homeowners the most. Or the policy concern with distributional effects may stem from a view of energy as a necessity; an increase in energy prices must be accommodated by reducing the consumption of other goods, a reduction which the poor can afford less well than the rich.

But this condition is not unique to energy, and there may be better avenues than energy policy for controlling the distribution of wealth. There does not appear to be even a simple rule for evaluating the distributional effects of energy policy. The aggregate data indicate that

Fig. 6. Energy Purchases versus Household Expenditures

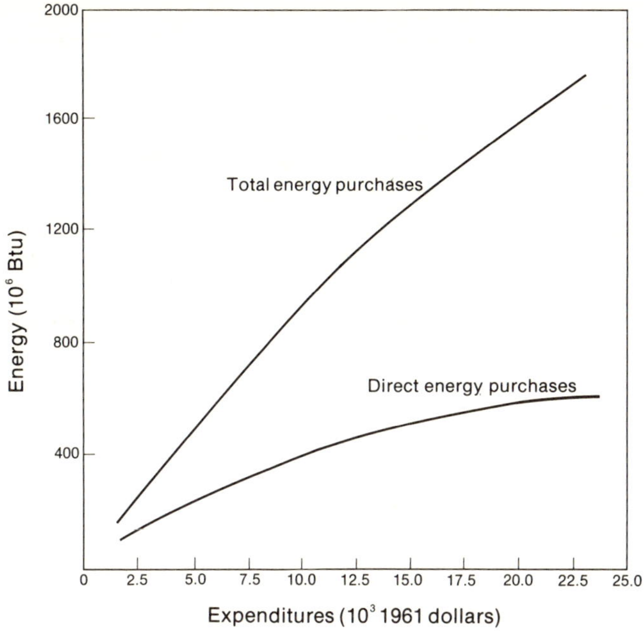

Source: R. Herendeen and J. Tanaka, "Energy Cost of Living," *Energy,* Vol. 1, No. 2, Pergamon Press, Great Britain, June 1976.

future discussions of the income effects must be related to specific energy policy proposals, and there is a need for better data and models to shed some light on this important dimension of energy policy.

Energy and Economic Growth: Summary

The historical link between energy and economic growth can be explained by the gradual decline in energy prices without adopting the philosophy of energy-economic determinism. In the short run, rapid rises in energy prices are difficult to manage, and there can be severe disruptions because of a sudden change in energy availability. But, in the long run, energy costs are a small component of the economy, and there is a substantial potential for adopting more energy-conserving modes of production and consumption; this flexibility can mitigate the aggregate economic impacts of a gradual increase in energy scarcity, whether the scarcity is imposed on us or adopted by us in exchange for other benefits, e.g., environmental improvement.

The loss in output is still significant, however, and it may disguise major shifts in the distribution of wealth, shifts that will not be accommodated easily. Energy scarcity is a problem, therefore, but not an insurmountable obstacle. The major difficulties are in moderating the short-run disruptions and compensating for the distributional effects: the journey is more important than the destination.*

This diagnosis of the energy-economic growth nexus is optimistic, perhaps excepting comparison with the views of the most enthusiastic conservationists; but it should not be interpreted as a signal to relax. Energy production is one of our most important single industries. Furthermore, the identification of the potential flexibility of energy utilization in the economy is not the same as the realization of that potential. The aggregate estimates of the elasticity summarizes but does not identify the essential details of the processes by which the energy substitutions will take place.

As a first step to achieving this conservation, the energy market must be rationalized, through deregulation and decontrol, so that energy prices reflect the marginal costs of energy production. The higher energy prices will provide the needed incentive; the aggregate analysis indicates that the opportunities are there for both energy conservation and economic growth.

---

*Lester Lave formulated this succinct characterization of the nature of the energy crisis, capturing the distinction between the ephemeral problem of resource exhaustion and the serious problem of adjusting to higher energy prices.

This essay is a merger and summary of the main themes developed in a series of related papers to which the reader is referred for further details: Energy Modeling Forum, *Energy and the Economy,* Vols. 1 and 2, Stanford University, Stanford, California, 1977; W. Hogan and A. Manne, "Energy-Economy Interactions: The Fable of the Elephant and the Rabbit?" in R. S. Pindyck, ed., *Advances in the Economics of Energy and Resources,* JAI Press, Greenwich, Conn., 1978; W. Hogan, "Energy and Economic Growth," in J. Sawhill, ed., *Energy Conservation and Public Policy,* Prentice Hall, N.Y., 1979; and W. Hogan, "Dimensions of Energy Demand," in *Selected Studies on Energy: Background Papers for "Energy: The Next Twenty Years,"* Resources for the Future, Washington, D.C., 1979.

Discussion

From the floor: One of your graphs shows the average cost of delivered energy in seven or eight countries in addition to the United States. Is it your contention that the prices here and in Canada will remain substantially lower than in the other OECD countries?

Hogan: They certainly have been in the past. The question is, what would bring the price of energy to the consumer here to levels more similar to those elsewhere? The only thing I can see is adoption of their policies, which put a very heavy tax on energy for private use—the end uses of energy. Obviously that is not popular from a political point of view. It is difficult to get even moderate taxes on gasoline through our Congress, to say nothing of the sort of taxes they pay in Italy or West Germany.

From the floor: Will decontrol bring more realistic pricing?

Hogan: Decontrol accounts for just a few cents a gallon. It is not very much. I think decontrol is a good idea; I favor it. But we should not delude ourselves into thinking that decontrolling energy prices in this country is going to raise prices to levels comparable to those in Western Europe. Nothing of the sort will happen. In fact, over a ten-year period, I think the changes implied by decontrol today will be swamped by what OPEC is going to do. That is a far more important and more uncertain element in predicting what is going to happen to oil prices.

From the floor: My question concerns a policy package that could accomplish investment and capital requirements for major expansions of energy supply. What role do you see for tax incentives in that package?

Hogan: The capability to provide the necessary investment is present in our economy; we just have to find the mechanisms to get it in the right places. As to financing options for energy investments, I think it is possible that some kinds of tax incentives might help, but I am not sure I would endorse them. Sam Schurr made a point I support: we should be certain to make only the energy investments that we find cost effective; we should not try to expand energy supply per se.

From the floor: The point has been made that we need an adaptive learning system that allows people to make choices. How do you see such a system allowing people to make choices?

Hogan: Decontrolling oil prices would be an important step along that way. We have now a labyrinthine system with all kinds of conflicting regulations that induce people to do things that are not in the national interest, and in some cases may not even be in their own interests, though they often cannot figure that out because the regulations are so complicated. If we could get rid of price controls, we would stop wasting energy on things that are not worth spending energy on at any price.

From the floor: Do you think gasoline rationing will improve people's opportunities to make choices?

Hogan: What it does is improve the opportunity for government to make choices for them, and I would not say the government has a very good track record in that regard. Most of the rationing plans that have been proposed are, to my mind, seriously deficient. Not that I have a better plan, but it is a fundamentally insoluble problem.

From the floor: How do you propose to teach people that they would be better off in the long run to pay larger initial costs for technologies that on a lifecycle basis would be more economical and also have the side benefit of reducing energy consumption?

Hogan: This is a common problem with small consumers. I imagine if energy prices continue to increase, people will naturally pay more attention to this sort of thing. We certain see a lot of inventive reactions in dealing with gasoline shortage problems. One solution would be to have the utilities finance the more energy-efficient technology for consumers—handle the financing, the servicing, provide the energy to run the more efficient (though more expensive) heat pump or refrigerator or whatever—and charge the consumer a monthly rate for the package. That might be the answer, but the problem needs a lot more work.

Chairman: The talks we have heard thus far in this conference seem to be lending credibility to the idea that the American economy can in the long run adjust to higher prices, but the problems are in the intermediate period: how to get the maximum advantage of private investment and the free market; what the role of government should be, since it is clear that it cannot be totally out of the picture; how to educate the voter who will then influence the Congress. As a professional educator, Dr. Hogan, would you care to say more about the process of getting the facts, alternatives, consequences, conclusions to the public?

Hogan: From a historical perspective, when you consider the magnitude of the price change and the importance of the energy sector, this is an unusual problem for our society to deal with, so maybe the response is not as slow as it might appear.

But there is another aspect to the situation, and that is the complexity of the problem. It has many different issues and they get confused in the discussion. The serious environmental problems tend to get mixed up with problems of national security, what's happening with OPEC oil, and our vulnerability in being dependent on the Middle East. Then there is the economic problem—higher prices and how to accommodate, what that will do to people's standard of living. Often proponents of a solution to one problem use arguments from another in order to make a point. So I think part of the educational approach is to separate the various issues and try first to decide what is important and what unimportant.

David Sternlight

Some Effects of Higher Oil Prices

Higher prices for oil encourage conservation, reduce demand, and stimulate supply. To permit the market to reach price levels at which supply and demand will balance, the federal government has decided to decontrol oil prices. Over the short- and medium-term, what are the likely effects of decontrol on inflation, on economic growth, and on employment?

Because many in this audience have already seen analyses of decontrol and its consequences during the course of the ongoing debate that led to the President's proposal, I shall precede my answer to that question with consideration of a related topic, having to do with the costs of some energy options that look promising, such as synthetics. What happens to the costs of coal and shale as world oil and energy prices rise? Are shale costs, for example, locked into oil prices so that shale oil cannot compete with oil in the future?

Costs of Energy Alternatives (Shale and Coal)

A pragmatic analysis was undertaken by Atlantic Richfield to examine the relationship of energy prices to inflation in energy project costs, both historically and in the future. The analysis began with "recipes" for

---

David Sternlight *is Chief Economist for Atlantic Richfield Company. Formerly Deputy Director of the U.S. Secretary of Commerce's Office of Policy Development, he has also directed economic planning for Litton Industries. Dr. Sternlight is a consultant to the United Nations and UNESCO. He is a member of the Council on Foreign Relations of the Economic Policy Council, United Nations Association of the U.S.A; the Workshop on Alternative Energy Strategies; and the M.I.T. World Coal Study.*

major projects, which developed project-related cost indexes from the major components of project costs. Both the Data Resources model of the U.S. economy and the Data Resources cost forecasting model (which reflects prices through the stages of processing in the economy) were used to stimulate project cost behavior without the 1974 OPEC price shock. These simulated cost estimates were then compared with project cost estimates that occurred historically. Finally, these costs were projected from 1978 to 1990, first under an assumption of constant real energy prices, and second under an assumption of a 3 percent "real" price rise each year. In both cases, inflation must be added to the real price assumptions to get actual price estimates for each year.

For a simple illustration, let us assume a project cost formula with only three variables. If a project uses (a) contract construction, (b) finished steel products, and (c) pumps and compressors, then we might construct a project cost index from three weights, all multiplied by the basic wholesale prices of the project's inputs. (The weights, of course, add to one.) This process translates inflation in wholesale prices into cost index increases for a specific project. The input indexes are taken from U.S. Government-published data, and the weights are based on the 1974 project cost mix, which seems to be fairly robust over time.

Following that formula, Table 1 shows some "recipes" for oil shale and for surface and underground coal. The weights for an oil shale (surface retorting) plant presently under development show, for example, that construction wages are about 40 percent of the costs and

Table 1. Weights by Project

Cost Components	Oil Shale	*Coal* Surface	*Coal* Under-Ground	*Annual Average Growth Rates* 1966-72	*Annual Average Growth Rates* 1972-78
Contract Construction Wages	**.40**	**.15**	**.20**	**7.7**	**6.0**
Finished Steel Products	**.30**	**.15**	**.20**	**4.7**	**11.7**
Pumps and Compressors	**.15**	**.15**	**.10**	**4.4**	**10.7**
Mining Machinery	**.06**	**—**	**.40**	**3.2**	**13.4**
Construction Machinery	**—**	**.50**	**—**	**4.5**	**10.8**
Industrial Commodities (+3%)	**.09**	**.05**	**.10**	**6.0**	**13.0**
	1.00	**1.00**	**1.00**		

Fig. 1. Ratio of Project Costs to World Oil Price

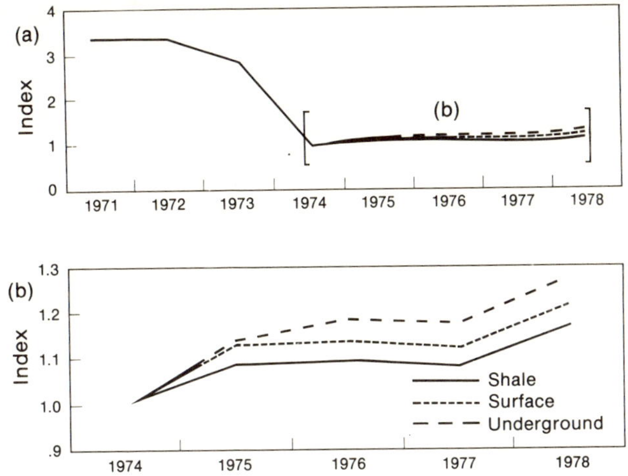

finished steel products about 30 percent. We also see the average annual growth rates for these different costs over time.

Some people have believed that these costs would always increase with world oil prices. Is this true? Figure 1(a) shows the ratio of project costs to world oil price. For a shale surface project, and for surface or underground coal, since 1974 the project costs seem to have moved almost in lockstep with world oil prices. But when in Figure 1(b) the representation for 1974-1978 is magnified, it becomes apparent that they have moved somewhat differently; the rates of growth and the slopes of the inflection points have been a little different. From 1975 to 1977 there has been some significant directional difference.

Between 1972 and 1977, world oil prices rose by 390 percent. Without OPEC we estimate they might have changed about 33 percent. Let us look at the results of a "no OPEC" scenario.

Table 2 shows cost component increases during that time, and Figure 2 shows project cost indexes. The dotted line is what the costs would have been without OPEC in the simulation, and the solid line is history. Without the OPEC price shock there would have been a 50 percent cost estimate increase for the shale plant over the period 1972 to 1977, compared to the 59 percent estimated cost increase ("history") with the OPEC shock included. This suggests that the cost increases in the shale estimates over that period were due to something other than the OPEC

Fig. 2. Project Costs, 1972-1977, with and without OPEC

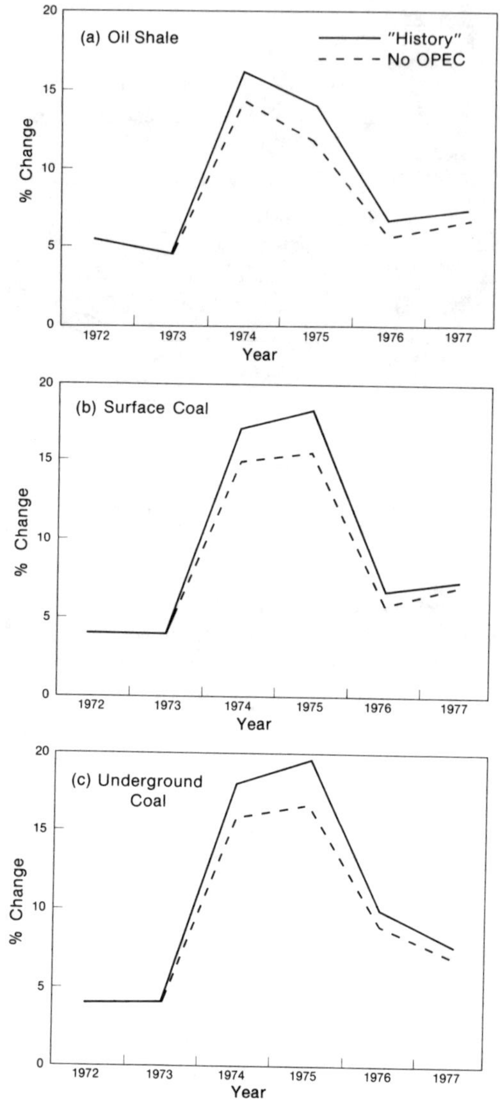

	% Change "History"	No OPEC	% Pt. Diff.
(a) Shale	59%	50%	9% Pt.
(b) Surface	66	57	9
(c) Underground	75	65	10

price rise. Similarly, surface coal costs would have increased 57 percent without OPEC compared to 66 percent actual; underground coal 65 percent compared to 75 percent. Thus most of the increases would likely have occurred without an OPEC price shock.

For 1974, the relative contribution to the inflation rate of the major shocks we experienced are shown in Table 3. Out of the 10.4 percent inflation rate only about 1.7 percentage points were due to the energy crisis. Another 1.5 were due to the agricultural price explosion, 1.2 due

Table 2. Historical Analysis of Impact of Oil Price Increase, 1972-1977

Cost Components	% Change		% Pt. Difference
	History	No OPEC	
Contract Construction Wages	**34%**	**29%**	**5% Pt.**
Finished Steel Products	**76**	**68**	**8**
Pumps and Compressors	**71**	**54**	**17**
Mining Machinery	**95**	**84**	**11**
Construction Machinery	**70**	**63**	**7**
Industrial Commodities (+3%)	**89**	**65**	**24**

Table 3. Impact of Shocks on Inflation Rate, 1974

Shock	Contribution to Inflation Rate
Energy Crisis	**1.7%**
Agricultural Price Explosion	**1.5**
Devaluation of the Dollar	**1.2**
Monetary Policies	**0.9**
Price Decontrol	**0.8**
Fiscal Policies	**0.3**
Total	**6.4%**
Base Rate	**4.0**
Rate of Inflation, 1974	**10.4%**

Source: Data Resources, Inc.

to devaluation of the dollar, almost 1 percent came about through monetary policies, .8 percent through lifting price controls, and a little bit from fiscal policies, for a total of 6.4 percent. Adding this to a base rate of about 4 percent gives the 10.4 percent rate of inflation in 1974.

Let us test that insight and see what it might mean for the future. Table 4 and Figure 3 show the results of a simulation from 1980 to 1990. In the base case we assume a percentage change in nominal world oil prices of 90 percent, essentially the rate of inflation, and in the higher oil price case we assume a 3 percent real price increase per year, or 153 percent change over the period, in nominal terms.

Table 4. Higher Oil Price Impact on Cost Components, 1980-1990

Cost Components	% Change		% Point Difference
	Base Case	Higher Oil Prices	
Contract Construction Wages	**128%**	**130%**	**2% Pt.**
Finished Steel Products	**117**	**118**	**1 Pt.**
Pumps and Compressors	**94**	**98**	**4**
Mining Machinery	**101**	**102**	**1**
Construction Machinery	**98**	**100**	**2**
Industrial Commodities (+3%)	**144**	**152**	**8**
Addendum: World Oil Price	**90%**	**153%**	**63% Pt.**

Table 4 shows the differences in individual wholesale price indexes. Contract construction wages differ by only 2 percentage points between the higher oil price case and the base case, 128 percent increase over the period, compared to 130 percent. The largest increase is in "industrial commodities plus 3 percent." There the difference is only 8 percentage points over the entire ten year period, compared to a 63 percentage point increase in oil prices between the base case and the higher oil price case over the period. Thus the wholesale price components of energy project costs seem to increase by only a small increment as a result of higher OPEC prices.

The effect on overall project costs can be seen in Figure 3. The annual rate of increase in the higher oil price case is within one- or two-tenths of a percentage point per year of the base case throughout the forecast

Fig. 3. Higher Oil Price Impact on Project Costs, 1980-1990

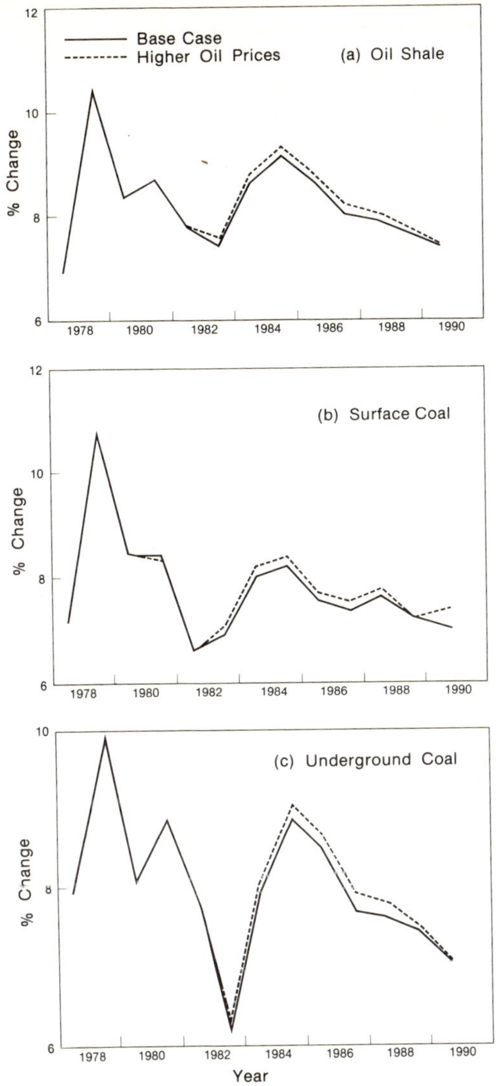

	% Change 1980-90		
	Base Case	Higher Oil Prices	% Pt. Diff.
(a) Shale	119%	121%	2% Pt.
(b) Surface	106	109	3
(c) Underground	112	114	2

interval for each project type with a little higher increase in underground coal costs because of the higher labor cost content. The cumulative impacts over the ten-year period are on the order of 2 and 3 percentage points, on a cumulative oil price rise of 63 percentage points.

This analysis suggests that shale, surface and underground coal and perhaps a number of other energy project costs are not locked to oil price rises. Rather, the behavior we have seen historically has been due to factors in the economy other than energy price rises. Agricultural price shocks, for example, were climate-induced. Thus, this analysis suggests that many alternative energy source costs are little influenced by OPEC price rises, and may be expected to become increasingly competitive with the cost of OPEC oil over time, other things being equal.

Macro-Economic Issues

Let us now examine the relationship between oil price decontrol in the short- to medium-run, and inflation and real economic growth. The analysis I will describe, which parallels our own findings, was made available by Frank Potter of the House Subcommittee on Energy and Power staff. It was performed by Chase Econometrics in early 1979.

The Chase Econometrics model was used to simulate the effects of decontrol using conservative assumptions; that is, the results reflect higher inflation and lower real growth than they would with more ambitious assumptions. For example, Chase assumed there would be no supply response from decontrol. They assumed that refiners would pass cost increases through dollar-for-dollar rather than absorbing some of those cost increases.

Their base case assumed domestic oil prices rising with inflation to 1985, and they compared several alternative policy options to the base case. Among those policy options examined was instant decontrol, which has deleterious effects on the economy that none of us would like to see. The Chase cases that are interesting consider decontrol over two years, with and without a severance tax on oil company revenues and with or without plowing that tax back into the general economy. These cases reflect current policy choices.

The theory behind Chase's work is cost push; as prices rise, inflation rises. It is not a monetarist model. In a monetarist view, the only thing that causes inflation is printing too much money. Monetarists believe that if the price of something rises, then people have less money to spend on something else and the economy balances out. That is not the Chase view, nor is it my own.

Table 5. Economic Impacts of Phased Decontrol During 1979-1981

	1980	1981	1983	1985
Consumer Price Index				
Base	225.9	240.1	269.3	303.1
Decontrol	226.4	241.0	270.4	303.8
% Difference	+.2	+.4	+.4	+.2
Real GNP (1972 $ billions)				
Base	1430.1	1482.1	1579.4	1686.9
Decontrol	1428.3	1478.3	1575.4	1683.8
Difference	−1.8	−3.7	−3.9	−3.2
Unemployment Rate				
Base	7.6	7.2	6.1	5.1
Decontrol	7.6	7.3	6.2	5.2
% Pt. Difference	0.0	+.1	+.2	+.1
Increased Producer Revenues ($ billions, nominal)	6.8	12.4	13.0	11.9
Average Increase in Product Price (cents/gallon, nominal $)	2.3	4.1	4.2	3.8

Note: Numbers may not add exactly due to rounding.

Let us examine the results of some of these analyses. Table 5 shows the economic impacts of phased decontrol from 1979 to 1981. I must warn you that these figures are not presented in the way to which economists are accustomed. We are used to looking at the percentage change from one quarter to the next at annual rates, or from one year to the next. Table 5 shows the *levels* of prices measured by the Consumer Price Index and the *level* of real GNP, and the differences between those levels under phased decontrol and the base case. The reason is that the percentage-change differences are infinitesimal. In 1980 under phased decontrol the Consumer Price Index might be two-tenths of a percentage point higher than it would be in the base case. It rises to four-tenths of a percentage point above the level of the base case, and finally settles down in 1985 at about two-tenths of a percentage point difference. These are cumulative differences. The difference in real GNP is about 1.8 billion dollars ($1972). That is a lot of money even though the percentage of GNP is an extremely small figure. The difference rises cumulatively to about 3.2 billion dollars ($1972) at the end of the period. The unemployment rates differ by about one- or two-tenths of a percentage point.

The point to note especially in Table 5 is the size of the increases in the Consumer Price Index—on the order of two-tenths to four-tenths of a

Table 6. Economic Impacts of Continuation of Present Policy Followed by Sudden Control after September 1981

	1980	1981	1983	1985
Consumer Price Index				
Base	**225.9**	**240.1**	**269.3**	**303.1**
Decontrol	**225.9**	**240.3**	**270.4**	**303.9**
% Difference	**0.0**	**+.1**	**+.4**	**+.3**
Real GNP (1972 $ billions)				
Base	**1430.1**	**1482.1**	**1579.4**	**1686.9**
Decontrol	**1430.1**	**1481.3**	**1575.0**	**1684.4**
Difference	**+0.0**	**−.8**	**−4.4**	**−2.6**
Unemployment Rate				
Base	**7.6**	**7.2**	**6.1**	**5.1**
Decontrol	**7.6**	**7.2**	**6.3**	**5.2**
% Pt. Difference	**0.0**	**0.0**	**+.2**	**+.1**
Increased Producer Revenues ($ billions, current)	**0.0**	**3.4**	**13.0**	**11.9**
Average Increase in Product Price (cents/gallon, nominal $)	**0.0**	**1.1**	**4.2**	**3.8**

Table 7. Economic Impacts of Phased Decontrol with Severance Tax and No Tax Rebate

	1980	1981	1983	1985
Consumer Price Index				
Base	**225.9**	**240.1**	**269.3**	**303.1**
Tax	**226.4**	**241.0**	**270.4**	**303.7**
% Difference	**+.2**	**+.4**	**+.4**	**+.2**
Real GNP (1972 $ billions)				
Base	**1430.1**	**1482.1**	**1579.4**	**1686.9**
Tax	**1428.0**	**1477.2**	**1573.6**	**1683.6**
Difference	**−2.2**	**−4.8**	**−5.7**	**−3.3**
Unemployment Rate				
Base	**7.6**	**7.2**	**6.1**	**5.1**
Tax	**7.6**	**7.3**	**6.3**	**5.2**
% Pt. Difference	**+.1**	**+.1**	**+.2**	**+.1**
Increased Producer Revenues ($ billions, nominal)	**3.95**	**8.3**	**10.9**	**11.9**
Severance Tax Revenues ($ billions, nominal)	**2.8**	**4.1**	**2.1**	**0.0**
Average Increase in Product Price (cents/gallon, nominal $)	**2.3**	**4.1**	**4.2**	**3.8**

Note: Numbers may not add exactly due to rounding.

percentage point difference. To put that in more conventional terms, the difference in the annual rate of change in the Consumer Price Index is about .2 percent: in the base case the CPI grows at about a 7.5 percent annual rate from 1978 to 1980 compared with a two-year decontrol growth of about 7.7 percent.

Table 6 shows the result of sudden decontrol in September 1981. It makes a difference of 4.4 billion dollars to real GNP in 1983; the economy recovers as revenues are subsequently reinvested. There is no unemployment effect in 1980 and 1981; there is a difference of about two-tenths of a percentage point in 1983 and 1984; the CPI rises three-tenths to four-tenths of a percentage point.

Table 7 shows phased decontrol with a severance tax and no tax rebate. The Treasury keeps the money; the Consumer Price Index rises four-tenths of a percentage point in 1981 and 1983. Real GNP is down, at worst in 1983 by 5.7 billion dollars ($1972).

Table 8, phased decontrol with a tax rebate, produces about a tenth of a percentage point more inflation in 1983 but the GNP growth is a little better. The differences are again very small. The main point is that in

Table 8. Economic Impacts of Phased Decontrol with Severance Tax and Tax Rebate

	1980	1981	1983	1985
Consumer Price Index				
Base	225.9	240.1	269.3	303.1
Tax	226.4	241.0	270.5	304.0
% Difference	+.2	+.4	+.5	+.3
Real GNP (1972 $ billions)				
Base	1430.1	1482.1	1579.4	1686.9
Tax	1429.0	1479.2	1574.9	1683.4
Difference	−1.2	−2.9	−4.1	−3.6
Unemployment Rate				
Base	7.6	7.2	6.1	5.1
Tax	7.6	7.2	6.2	5.2
% Pt. Difference	0.0	+.1	+.1	+.1
Increased Producer Revenues ($ billions, nominal)	3.95	8.3	10.9	11.9
Severance Tax Revenues ($ billions, nominal)	2.8	4.1	2.1	0.0
Average Increase in Product Price (cents/gallon, nominal $)	2.3	4.1	4.2	3.8

Note: Numbers may not add exactly due to rounding.

terms of the effect on real growth and prices, although the numbers are large, the case differences in percentage terms are small—small enough, in fact, to be within the normal error tolerances of economic forecasts and the econometric models themselves.

Figure 4 presents the real GNP in the three cases graphically. The bottom line is sudden decontrol in '79. There is a drop in real GNP at that point, while with phased decontrol GNP drops rather more slowly and then gradually rises as the effects of reinvesting the revenues find their way through the economy, starting in about 1982. (The graphs are in billions of constant 1972 dollars, not percentages.)

Fig. 4. Effect of Decontrol on Gross National Product

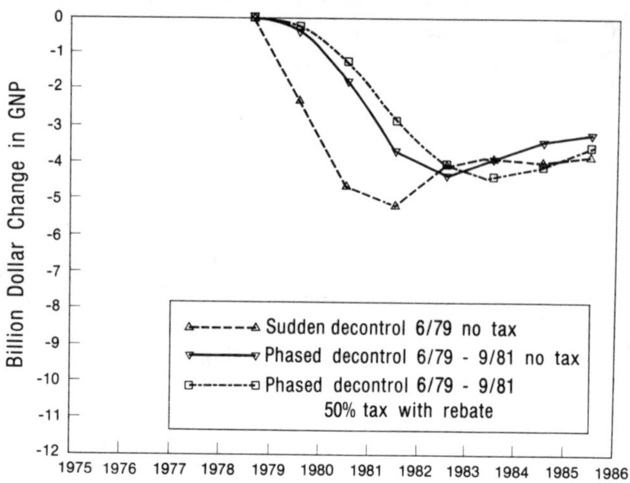

Figure 5 shows the Consumer Price Index, which rises with sudden decontrol, although the increment is about half a percentage point, and then starts tapering off. Under phased decontrol it is less dramatic. In the phased decontrol, no tax case, the initial effects of a bit more inflation and a bit less growth in the first year or two are compensated for toward the end of the period with a bit less inflation and a bit more growth, compared with the phased decontrol with tax or tax and plowback. In the phased decontrol, no tax case, the revenues are reinvested in the industrial sector of the economy, while the other cases have the revenues partly invested in the consumption sector of the economy and in services, when the Treasury spends the money.

The unemployment rate changes are shown in Figure 6.

Fig. 5. Effect of Decontrol on Consumer Price Index

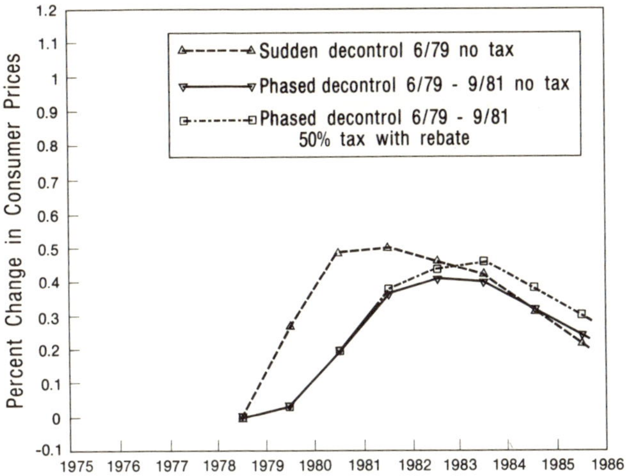

Fig. 6. Effect of Decontrol on Unemployment

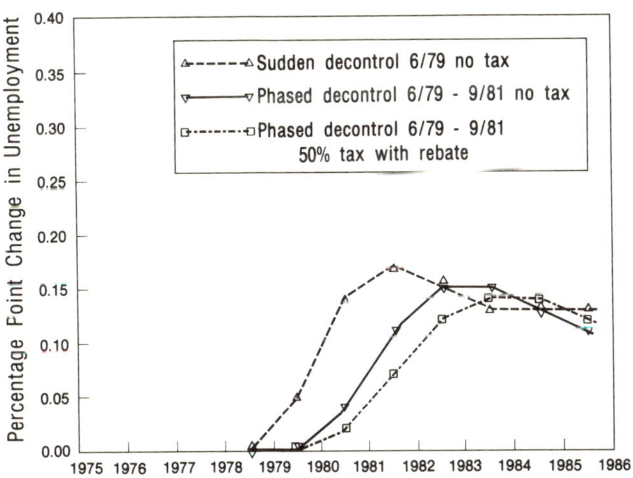

Conclusion

This paper provides directional indication that (1) in the short- to medium-term, the effects of two-year phased decontrol on inflation and on real growth are minor and damp out fairly quickly; and (2) higher oil prices do not make alternative energy supplies substantially more costly than now. The outlook is thus for abundance at higher oil prices, without severe inflationary consequences or any of the other deleterious effects that some feared as a result of the higher prices. However, none of these analyses assumes a new catastrophe that would cause a severe interruption in supplies; the hope is that the emergency petroleum reserve and related policies might handle such issues.

Discussion

Chairman: In some respects the conclusions that you have drawn from these econometric modeling studies would be startling to the American public, and I am wondering whether Atlantic Richfield is going to engage in a major publicity program with regard to these results?

Sternlight: It is not surprising that some of the results are startling because there have been developments in the past that seem to contradict them, such as sharp OPEC price rises, accompanied by adverse effects on the economy. As I have shown, those effects were in part due to OPEC price rises and in part due to other factors. The American public has become unable to distinguish, perhaps because of the treatment in the media, between higher OPEC prices and higher domestic prices. When OPEC puts the prices up, OPEC collects the money and to some extent that money comes back to the United States in the form of purchases of real goods and services. To a larger extent, however, that money comes back in financial investments which really do little more than relieve some pressures in the money market, if those pressures are present.

In contrast, higher domestic prices come back in the form of wages, salaries, profits and dividends, all of which go back into the American economy and further stimulate investment and growth in jobs. Thus the effects of higher domestic prices are quite different on the economy than the effect of higher OPEC prices. This lesson has not yet been sufficiently transmitted to the public. My experience is that when one takes the time with media people to sit down in a relaxed way and discuss what is going on and what the cause-and-effect relationships seem to be, they are quite willing to listen and to learn. We can see the result in the press, which seems to be ahead of the television networks in understanding these issues. It is a learning process, and what the oil industry has to do is to enhance that process.

From the floor: As a cautious representative of an oil company, you may be understating the case somewhat. Theory would tell you that if you are buying oil for $16 and selling it for $12 that you are losing money and that decontrol would actually result in an increase to the real GNP. The problem in these models is, first, they have a very difficult time in capturing the short-run adjustments in energy use, and second, there are a lot of assumptions embedded that are made by the modellers about how the fiscal and monetary policy varies during the transition. How much of what appears there is error in the models or pessimism on the part of the analyst that the country will not be able to manage the transition?

Sternlight: You have pointed out two issues. The first is that many people distrust oil companies, so a company tends to be cautious about presenting its case, preferring independent calculations when these are available. You perhaps noticed my own reluctance to present the analysis we at Atlantic Richfield did of the effects of decontrol using the DRI model, which was essentially identical to the analysis that the Energy and Power Subcommittee had done using the Chase model. When things do not quite work out the way expected, if the case has been overstated, people are sure that it was out of mendacity and not because it is one versus another reasoned analysis.

The second point is on the character of the models of the analysis. The models tend to be very cautious about supply elasticity. The Chase study does not assume any increase in supply. The DRI model calculations made at Atlantic Richfield were more optimistic than the Chase calculations in respect to the balance of payments because we used our best judgment about increased supplies and found that decontrol, after a lag, produced significantly increased supplies. This led to backing out imports, on a barrel for barrel basis, which led to an improvement in the U.S. balance of payments. This led to a further reduction in inflation; when the U.S. dollar is weak because the United States is running a balance of payments deficit, foreign goods that are competitive with U.S. goods cost more and provide a price umbrella, allowing sellers of competing U.S. goods to charge more at home than they otherwise could. If foreign cars become more expensive relative to American cars, and demand is strong, that allows American car manufacturers to raise prices higher than they might otherwise. The chain of events starting from increased supply through backing out imports through improving our balance of payments adds an additional element of inflation reduction: The increased investment also improves our investment-to-GNP ratio which further strengthens the economy.

To explain this in detail, and to know the values of the parameters well enough to defend them, is not always easy and so one tries to present conservative estimates. One analysis that Atlantic Richfield did reflected all of these interactions. The net post-decontrol position of the economy five years later was actually better in terms of real growth of GNP and lower inflation than the alternative. The Chase runs, done on a more conservative basis, do not have that result.

From the floor: What did you do about government economic policies?

Sternlight: The model was driven by a series of assumptions about tax policy, government fiscal policy, and the rate at which it was possible to deal with embedded inflation in the economy, which are current conventional wisdom; similar assumptions are made by DRI, Chase, and Wharton. In addition a cylical assumption was made that there would be a slowdown late in 1979 and that the economy would then resume moderate growth in mid to late 1980. That cyclical assumption caused some of the dynamics you saw.

From the floor: The United States imports oil in forms other than liquid oil from Arab countries. We import huge stocks of Toyotas and Datsuns that take oil to manufacture. Have you considered the increased imports of Toyotas and Datsuns that would result from higher domestic oil prices and thus higher U.S. car prices?

Sternlight: This was considered within the structure of the econometric models used, which contain prices for imports as well as domestic commodities of all kinds. Those price indexes are driven, in part, by energy prices. When energy prices change, prices throughout the economy change at an aggregated level in these models. This includes import and export prices.

From the floor: Will not decontrol raise prices at first, and reduce growth?

Sternlight: In the short term, decontrol does raise prices and probably would, to some extent, reduce real GNP. After about eighteen months, GNP would rise and inflation fall to at least offset these short-run effects. It depends on things like Federal Reserve policy, which might swamp the relatively small effects from decontrol we saw here.

From the floor: Why does growth go down at first, if there is more energy investment?

Sternlight: There are time lags in the process. With phased decontrol, prices rise gradually. This causes some increase in investment right away; it takes time for those commitments to work their way through to purchases of equipment and payments of wages and salaries which then create further demands in the economy. Thus the price rise causes an initial negative effect on real growth, followed by a positive effect as the working-through process gets going.

Chairman: No doubt a good many people here have had need for and have benefited from sophisticated econometric modeling that takes into account a great number of factors and gives a wide range of assumptions. The important conclusion is that in the near term, higher prices are not going to hurt because one can accommodate to them. In the short term, except for *sudden* decontrol, there is not much effect on growth or unemployment from decontrol. The amount we are talking about here is comparable to that resulting from errors in the models or from overlooking some factors, or from making slightly different assumptions about details of government tax policy, any of which would more than compensate for the initial negative effects. The startling thing is that decontrol does not show any substantial negative consequences, whereas the American public, and perhaps much of the media, seem to have the impression that the increased prices that decontrol allows have a tremendous impact. The reason why certain leading members of Congress are opposed to decontrol is because they anticipate that their constituencies will feel that they are selling America and its people down the river.

Sternlight: I would not put it quite that strongly. Within the range of the models' errors, there is a significant difference between phased decontrol, and no decontrol followed by sudden decontrol in 1981. The difference is that during the three to six months after sudden decontrol, the effects are much sharper. Over the interval the effects average out, but in the immediate period the effects are much sharper. If there is no decontrol now, during the period through 1981 OPEC prices will rise yet further than they would otherwise do. Domestic prices will rise perhaps only by inflation. The difference between OPEC and domestic prices by 1981 will be much greater than today. The shock the economy would have to absorb when the controls legislation expires would thus be higher.

Second, although changes in growth and inflation are very small compared to changes associated with business cycle developments, I would not want to be too callous about that. We must not neglect the social consequences of these things. We must make some compensating adjustments; perhaps income supplements or something of that sort for poor people. But you are quite correct—we are not going to damage the U.S. economy in any significant way by decontrolling oil prices.

Henry R. Linden

The Role of Gas in America's Energy Future

I have long been an exponent of the proposition that mankind's astounding population growth, social and economic progress, and general improvement in well-being since the Industrial Revolution have been due to the growing substitution of inanimate energy forms for human labor. In particular, the abolition of serfdom, slavery, and child labor in industrialized countries and most recently, the emancipation of women in the most affluent and enlightened societies clearly have been the result of the use of labor-saving technology fueled by seemingly inexhaustible energy supplies provided by nature in very convenient form.

The fundamental meaning of these momentous events for the vast majority of those living in less developed countries should not be obscured by the relatively minor differences in energy consumption levels and related economic, social, and cultural factors that differentiate the United States and, to a lesser extent, Canada from Western Europe, Australia, and Japan, or even from much of the Soviet bloc and other highly industrialized areas. Most in these advanced societies share

---

Henry R. Linden *is President and a director of Gas Research Institute. As an energy research manager and energy policy analyst for 35 years, he has lectured and published widely and holds many United States and foreign patents in fuel technology. He also teaches chemical and gas engineering at the Illinois Institute of Technology, serves on a number of industrial and public interest boards and government advisory bodies, and is a member of the National Academy of Engineering.*

the desire to achieve the unprecedented level of egalitarianism coupled with affluence and personal freedom that has been the North American ideal.

In addition to energy abundance, a key factor in the evolution of this egalitarian society has been physical mobility and its by-products—political freedom and social and economic mobility all within a framework of surprising cultural diversity. This exceptional mobility of North American society has been due in great measure to the availability of low-cost personal transport fueled by cheap gasoline. Thus, it is no surprise that those who disapprove of some major aspect of this society, reserve special scorn for the private automobile, while most who still strive to achieve its goals react rather strongly to threats of restriction of personal mobility.

In addressing these issues in a period that could easily become one of increasingly scarce and costly energy supplies, and resulting social, economic, and political upheaval, and perhaps even "resource wars," it is important to consider the role of leadership. I believe that the most important task for leadership is to state forcefully and believe strongly that energy use is good, not that energy use is bad, and to act in conformance with this proposition. As I have tried to outline only briefly, it is just not true that energy use is bad. All the historic evidence shows that the overwhelming majority of so-called "civilized" mankind prior to 1750 was miserable—and ruled and exploited by a very small upper crust that had extremely high per capita energy consumption when all of the human and animal labor required to support them is factored in. Perhaps the highest per capita energy consumption in human history was the Roman court which, for example, got its "air conditioning" via a bunch of elephants driven by God knows how many men going back and forth between Rome and the Alps during the summer to haul in snow. This historical record of social and economic injustice and human misery as a result of energy scarcity should be remembered when considering the topic of "an era of abundant but higher priced energy" in America—which is the theme of this conference. My purpose here is to address the role of natural gas in such an era.

Natural gas has been a central factor in U.S. energy supply. It reached a market share of fully one-third of total primary energy consumption in the early 1970s, and still provides 25 percent today in spite of many years of decline in proved domestic reserves. My basic premise is that gas from conventional and unconventional natural sources and from a wide variety of synthetic sources could and should maintain at least its current position in meeting the nation's energy supply. This translates into total pipeline-quality gas needs of nearly 25 trillion cubic feet (or quadrillion Btu—quads for short) in 1985 and about 30 trillion cubic feet or quads in 2000, far above most projections by private and government forecasters.

However, this negative attitude toward gas is changing rapidly. The strongest driving force behind this change is growing awareness that one of the best means for containing U.S. oil import requirements in the near term would be policies that ensure a continuing contribution of at least the current 20 quads of natural gas from existing sources, namely, the lower 48 United States, Canada, and on-going imports of liquefied natural gas (LNG).

There is also growing awareness that, over the midterm, the economically and environmentally most acceptable way to maintain adequate energy supplies without relying on ever-increasing and monetarily unmanageable oil imports from possibly insecure sources must include full development of the potentially huge unconventional natural gas resources and of the demonstrably huge synthetic gas-from-coal capability.

In the long term, renewable sources of methane (the main constituent of natural and synthetic pipeline-quality gas) derived from marine and, possibly, land-based biomass, will surely play an increasingly significant role. It goes without saying that conservation and solar-thermal augmentation are an integral part of this scenario which I call "the gas option" and, in fact, the very nature of pipeline-quality gas supply and utilization is peculiarly suited to enhance the contributions from these essential components of the solution to the energy problem.

So my view of the energy future of the United States is certainly influenced by my conviction that maintaining a substantial market share for gas is going to be very beneficial to energy consumers and to the national interest as a whole. With that as a starting point, let us examine some of the background of the problem and the possible solutions.

Role of Gas in Energy Policy

For the past two decades, three-quarters of the nation's energy has been supplied by liquid and gaseous fuels. Most of the increase in energy demand has been satisfied with imported oil for over half of this period, while the share of coal in the primary energy market has remained at 17-20 percent. Thus, the policy of subsidizing imported oil has been all too successful. This is especially true in the industrial boiler market where gas and oil are totally exchangeable and where, in many cases, conversion to coal is feasible.

In 1978, the importation of an additional million barrels per day of oil was thought to be due to displacement of gas with oil in the industrial market due at least in part to excessive restrictions on gas use. Thanks to the beneficial impact of the Natural Gas Policy Act of 1978 on interstate gas supply and the constructive manner in which the National Energy Act (NEA) and Fuel Use Act (FUA) of 1978 are being implemented, this trend has now been reversed.

Quite clearly, any near-term solution of the oil import problem should include the following initiatives:

1. Maximize domestic oil and gas production and imports from relatively secure and, preferably, soft currency sources.
2. Use no liquid fuels where conventional natural, unconventional natural, or synthetic gas can do the job, because in most stationary heat energy uses liquid and gaseous fuels are totally substitutable.
3. Use conventional petroleum liquids primarily as a source of transport fuels.

Remember, because oil and all forms of natural gas and its synthetic substitutes are interchangeable energy sources in most stationary applications, every annual increment of pipeline-quality gas supply of two trillion cubic feet (equivalent to roughly 10 percent of current consumption) replaces roughly one million barrels per day of oil imports.

In the longer term, we will have to come to grips with the illogic of prematurely making radical changes in the infrastructure of energy supply and use merely because conventional oil and gas are running short. In justifying such radical change economically, advocates generally overlook the horrendous costs of converting to new energy forms the transportation, distribution, and utilization systems largely dependent on the direct use of gaseous and liquid hydrocarbon fuels.

While sitting in the midst of abundant source materials for oil and natural gas substitutes—coal, oil shale, unconventional natural gas, organic wastes, biomass, and a wide variety of other heretofore underdeveloped energy resources—it would make a lot more sense economically, environmentally, and socially to provide the missing links between these resources and their effective use within the existing infrastructure until wholesale conversion to new energy forms really becomes necessary. These missing links are, of course: (1) better technologies to produce heretofore marginal resources, and (2) a synthetic and biomass fuels industry.

Advantages of Gas as an Energy Source

In preserving the existing energy infrastructure as long as possible during the protracted, painful, and inevitable transition to inexhaustible or renewable energy sources and, more broadly, to a steady-state resource economy, pipeline-quality gas is destined to play a pivotal role. Gas is, environmentally, the most desirable primary energy form and is available to most end-users today at prices that are competitive with electricity and distillate fuel oil, which have equivalent form value. Further, on the basis of existing legislation and anticipated regulatory policies, gas promises to remain the most economical fuel in most energy markets for the foreseeable future.

Among the various primary energy sources, gas has the greatest flexibility, being already a refined product, suitable for all stationary heat energy uses and for many petrochemical uses, and storable relatively cheaply in large quantities in underground reservoirs.

The in-place transmission, storage and distribution system for pipeline-quality gas (that is, gas in the 900 to 1100 Btu cubic foot heating value range), built at a cost of $50 billion, linked into a national grid and literally irreplaceable today, is available to serve most residential, commercial, and industrial heat energy markets. It has sufficient additional capacity to accept substantial quantities of supplemental sources of gas at a relatively low incremental cost. (It is seldom realized that the total energy capacity of the existing U.S. underground gas storage system is larger than the elusive goal of a one billion barrel strategic oil storage capability, and that the equivalent of one half billion barrels of oil is actually deliverable during the winter months from gas storage.)

One reason for optimism for sufficient supplies of natural and supplemental gas to maintain at least the current 25 percent share of the U.S. energy market is growing evidence of the substantial supply elasticity of natural gas at prices competitive with energy products of equivalent form value.

The widely projected decline in the supply of natural gas, therefore, need not occur until after the turn of the century, because the quantities of conventional natural gas recoverable at these price levels is substantially higher than originally assumed and can probably be doubled by the inclusion of marginal and unconventional sources of natural gas which have been considered uneconomical. In fact, the domestic resource base of naturally occurring hydrocarbon materials is substantially greater for gas than for liquids at equivalent price levels.

Thus, on the basis of even a conservative assessment of available data, proved and potential reserves of natural gas recoverable from conventional and the more promising unconventional resources at acceptable price levels are equivalent to at least a 40-year supply at current rates of consumption.

A similar favorable situation in regard to gas exists in the area of synthetics. Synthetic gas from coal, oil shale, peat, biomass, and other carbon-containing sources could provide very substantial additional supplies at prices competitive with other synthetic fuels options. Not only do the projected costs appear to be less than for synthetic liquids, the technology for synthetics production is more advanced for gas.

Further, even on a delivered marginal cost basis, synthetic pipeline gas from coal apparently can compete with coal-based electricity in much of the residential space heating market. Synthetics may play a useful role in power generation for certain retrofit applications and in new systems designed specifically around, say, a coal gasification step to optimize overall performance. However, for heat energy needs in general, synthetic pipeline-quality gas and other premium synthetics used as such have a basic cost advantage over electricity produced from synthetics.

U.S. Options in Meeting Energy Supply Problems

Three issues dominate any consideration of future energy supply:

1. Between now and perhaps as long as the turn of the century while world oil resources are still adequate, the threat of oil embargoes and the likelihood of frequent production curtailments and of underdevelopment of productive capacity in support of political and cartel pricing objectives inimical to the interests of major oil-importing nations.

2. Beginning sometime around the turn of the century, the likelihood of actual world oil shortages and, well in advance of that time, totally unmanageable increases in world oil prices which threaten not only U.S. economic and social stability, but also international stability, with the growing possibility of "resource wars."

3. Over the entire period under consideration, the likelihood of growing trade deficits, international monetary instability, and rapidly

increasing energy costs for any nation heavily and increasingly dependent on oil imports.

On the basis of these considerations, a broad consensus has been reached in the United States on the need for greater energy self-sufficiency. Strategically, economically, politically, and socially, a lack of action to forestall the tragic consequences of these events by the military protector and economic leader of the non-communist world would be unthinkable.

Due notice should, of course, be taken of the brighter aspects of the world oil and gas outlook. The laws of economics have not been suspended. On the supply side, the classic effects of higher real prices aided by improving exploration and production technology are already apparent. Large new reserves are being added to world hydrocarbon supplies, including by our immediate neighbors. On the demand side, the twin forces of price elasticity and reduced economic growth have made their matching impact.

However, no responsible leader in the world energy community would suggest that a return to a business-as-usual approach can be justified under today's constraints on development of new energy sources. A lead time of 25 or even 50 years to replace much of the world's conventional hydrocarbon fuel supply and take care of demand growth is much too short for complacency.

In spite of the slow response to the energy problem, which surfaced first as a natural gas supply shortage in 1968, the United States has at its disposal a much wider range of options than do Western Europe and Japan. One of these options, attainment of virtual energy self-sufficiency in the next 20-25 years without a major economic set-back, is, in fact, credible only for the United States among the large energy importers. This is due to a fortunate combination of:

1. A still substantial capability for conventional domestic oil and natural gas production;
2. Large marginal hydrocarbon resources that can be developed at acceptable costs;
3. A very sizeable near- and mid-term energy conservation potential with minimal prospective economic and social impacts;
4. Very large and relatively cheaply recoverable coal and oil shale resources;
5. Acceptable existing and rapidly developing new technology for synthetic fuels production;

6. A nuclear power industry backed by adequate nuclear fuel resources still capable of major expansion in spite of recent severe setbacks;

7. A very strong energy R&D establishment, which is rapidly developing options for acceptable means of long-term energy conservation and for economically and environmentally acceptable new sources of domestic energy supply.

Potential Impact of Reduced U.S. Energy Supply

A program of increased energy self-sufficiency for the United States obviously must incorporate all available cost-effective and socially acceptable means to conserve energy. Unfortunately, the impact of reduced energy consumption on various sectors of the economy and on a wide range of important social values, either through conscious conservation efforts or as a result of internally or externally imposed limitations on energy supply, is not well understood. Therefore, prejudgments on the best paths and possible magnitudes of energy conservation are sources of concern.

The historic behavior of energy consumption in the United States from 1947 through 1977 can be explained with acceptable accuracy using only two variables: the deflated price index for fossil fuels and power, and the millions of people in the civilian labor force actually employed. Roughly, energy consumption has increased (or decreased) at twice the rate of the increase (or decrease) in civilian employment. On the other hand, for every 10 percent increase (or decrease) in deflated energy price index, energy consumption has decreased (or increased) by 2 percent.

Although it is questionable if these numerical relationships will hold in the future under the radically changed circumstances, they would indicate that a reasonable level of primary energy consumption in 1985 would be somewhere between 90 and 100 quads. For the year 2000, the range would be 120 and 130 quads. Although the range for 2000 is on the high side based on current thinking, these projections are quite similar to those in National Energy Plans I and II. They are also substantially below those made prior to the oil embargo because they reflect the much higher energy prices and the much lower economic growth rates compared to those anticipated prior to the oil embargo.

Inasmuch as energy policy is tightly linked to economic and social policy, it cannot be treated as a separate entity. Historically, per capita disposable personal income in constant dollars has just about perfectly tracked per capita energy consumption. Thus, any policy that has as its objective continued increase in real disposable personal income thanks to increased productivity, a high level of employment, and control of increases in the real cost of energy, will result in higher energy consumption than a policy willing to accept less than optimal energy use in the economy. Under-investment in energy and substitution of labor for energy can only lead to declines in productivity and a decrease in real disposable income.

Thus, the goal of a primary energy demand growth rate of no more than 2 percent annually from 1985 on involves a certain degree of risk. This risk to the future social and economic well-being of the nation will be greatly increased if major energy policy initiatives are based on inadequate or erroneous analyses of their potential merit. A major flaw in past U.S. energy policy initiatives that will be considered here is a diminishing role envisioned for pipeline-quality gas. Obviously, the U.S. energy future will be radically different if gas maintains its roughly 25 percent current share of the primary energy market than if it is displaced by other energy forms.

The Myth of Gas Scarcity

To avoid repetition of past mistakes, it is important to review public perception of the role of gas, both natural and synthetic, in the national energy supply. The most serious of the erroneous premises of the past has been the assumption that the best way to handle the natural gas shortage is to restrict the markets for gas in favor of other energy forms such as liquid fuels, coal, and electricity, despite the fact that potentially abundant natural and synthetic sources of gas frequently offer a more economic and environmentally more acceptable fuel.

Restricting the market for gas because it is "scarce" leads to the self-fulfilling prophecy that there is enough gas to meet demand and that new gas supply initiatives are not needed. The question then is, what replaces gas that has been restricted from serving its traditional markets?

In the past, this has been primarily electricity for the residential and commercial markets, and imported oil in the industrial market. This is clearly undesirable from just about any viewpoint, be it long-term consumer cost, environmental impact, or the broad national interest. Market restrictions lead to such absurdities as the recent "gas bubble," which was simply an oversupply created by a variety of government interventions.

Although the Department of Energy has withdrawn the Market Oriented Program Planning Study (MOPPS), conducted largely under the auspices of the Energy Research Development Administration (ERDA), this study contained a wealth of extremely useful information and represented a welcome departure from intuitive decision-making. However, it also projected an oversupply of gas in 1985 simultaneously with a large deficiency of oil.

This highly illogical expectation was accepted in spite of the previously noted substitutability of oil and gas in nearly all stationary uses, and the high form value of gas. Moreover, the MOPPS study itself indicated that, at equivalent wellhead prices (i.e., $4 per thousand cubic feet or million Btu of gas and $24 per barrel of crude oil, both in 1975 dollars), more conventional natural gas should be recoverable in the lower 48 states than conventional crude oil, and that there are vast additional quantities of unconventional natural gas that should be recoverable at these competitive prices.

Coal Gas versus Synthetic Liquid Fuels and Coal-Based Electricity

The basic flaw in the logic of arbitrarily restricting gas markets is even more evident when the prices of synthetic fuels are compared. Substitute natural gas (SNG) from coal shows a definite cost advantage over other synthetic fuels in the same stage of development. According to several government studies, high-Btu gas from first generation plants is the least expensive synthetic fuel transportable by pipeline that is available from a commercial process.

Second-generation process SNG from coal shows an even greater cost advantage when compared with the refined liquid fuels derivable from coal. While crude shale oil is in about the same price range, it is not directly comparable because it also needs considerable refining to convert it into an acceptable liquid fuel.

A second widely-accepted premise has been that electricity is the preferred source of residential/commercial energy needs in the future. It is, of course, in the consumer interest to use electricity where it can provide competitive or superior service, and these uses are expected to provide continuing demand growth somewhat higher than that of other energy forms. However, to use electricity where other energy forms can do the job at lower cost and with less environmental impact is illogical.

Although the relative merits of coal-by-wire versus coal-by-pipeline are still being vigorously debated, a number of studies have shown that, except in limited regions especially favorable to electric heat pump performance, the annual cost of residential heating is higher for electricity from a conventional power plant with stack-gas scrubbers than for gas from a Lurgi coal gasification SNG plant. Thus, the common assumption that the electricity option is automatically preferable needs re-examination in the light of economic and environmental realities. Moreover, if gas can be produced from coal competitively, it needs to be no more "scarce" than electricity.

In spite of these facts, federal energy planners in the past have questioned the long-term role of gas energy. This was a major criticism of NEP-I, as was its lack of specifics for increasing other forms of energy supply. The second National Energy Plan (NEP-II) corrected some of these deficiencies, and a very substantial future role is now envisioned for gas. However, the potential contributions of coal gas are still understated, as are those of unconventional natural gas and imports.

In the synthetic fuels area, the reason for the low expectations for gas is DOE's strategy to lend very substantial budget support for production of synthetic liquids and solids from coal, while the only new synthetic gas supply initiative has been DOE support before the Federal Energy Regulatory Commission for partial rate-payer financing of the construction of commercial Lurgi plants. Such plants could provide a clean, refined, pumpable fuel at a cost competitive with processes based on pilot-plant scale liquefaction technology. DOE does not provide support for second generation high-Btu gas from coal demonstration in any way comparable to its support of liquefaction or solvent refining demos, even though the leading second generation gasification processes are in at least as advanced a state of development, and promise a substantial cost advantage over generally lower form value synthetic liquid fuels.

Current U.S. Energy Policy and the Role of Gas

Now that most of the new energy legislation is in place, and upsets of growing frequency and severity in the world oil situation look increasingly probable, it is time to assess whether the United States will actually undergo any major change in real (as contrasted with nominal) energy policy. The major unknowns making such an assessment difficult continue to be:

1. The degree of public patience with predictable and, in fact, inevitable "energy crises" in the midst of an abundance of supply options, including options for ample supplies of motor gasoline at costs no greater than what the thriving economies of Western Europe and Japan pay now.

2. The public acceptance of the new domestic oil pricing policy, which is moving toward early equalization with world prices through a process of phased deregulation accompanied by substantially higher margins for producers even under the various proposed "windfall profits" taxation schemes.

3. As a corollary to (2), the willingness and ability of the political process to maintain long-term support of new oil and gas supply initiatives by private industry at substantially higher costs, rather than react to recurrent supply crises with demand restraint and price control measures.

4. The degree to which conventional natural gas from the lower 48 states, Alaska, Canada, Mexico, and from overseas in the form of LNG and, perhaps, unconventional natural gas and coal gas will be used to reduce oil imports from hard-currency or insecure sources through vigorous federal action that strengthens, modifies, or even reverses nominal existing policies on industrial and power plant use of pipeline-quality gas.

5. The likelihood of Mexico joining the growing framework of mutually supportive energy trade policies of the United States and Canada in return for financial and technical assistance on politically acceptable terms, thereby allowing optimal development and marketing of Mexican hydrocarbon resources with an attendant easing in the prospective hydrocarbon deficit and associated monetary and strategic problems of the northern half of the Western Hemisphere.

Among the reasonably well-defined trends that will affect U.S. energy policy during the foreseeable future are:

1. Continued U.S. oil import needs which, in conjunction with the needs of the other major oil importers, will frequently exceed available supplies because of voluntary or involuntary limitations on production by the major exporters below attainable levels.

2. Continued discrimination against "long-haul" LNG imports vis-à-vis oil imports from the same sources, in spite of many well-substantiated and rational arguments against such a policy.

3. Continuation of a rate of orders for new nuclear power plants below the subsistence level for a viable U.S. nuclear industry in response to widespread public opposition; local moratoriums by state regulatory bodies; and loss of commitment to the near-term nuclear option by electric utilities because (a) rapidly rising costs and indefinite delays have severely reduced or eliminated most advantages over

coal-fired plants, (b) the supply and price of enriched uranium and the disposal of nuclear waste have become fraught with intolerable uncertainties, and (c), perhaps most important, because of the repercussions of the accident at the Three Mile Island power station.

4. With the exception of central power station use, no spectacular increase in the substitution of coal for oil and natural gas as an industrial fuel, because of logistic and environmental problems and unfavorable economics when the much higher investment costs of coal-burning facilities are factored in.

5. A sharply reduced annual rate of increase in electric power demand compared to the 6 to 7 percent or so, predicted only a few years ago, due to cost-induced conservation, a much slower pace of electrification, and the likely long-term slowdown in economic growth.

6. Continuation of a relatively low rate of growth of primary energy consumption compared to the 4 percent annual rate typical of the period from World War II until the embargo and, probably, even compared to the 3 percent long-term U.S. historical rate, again due to a slowdown in economic expansion and to cost-induced conservation enhanced by mandatory requirements such as in the automotive transportation sector.

7. Small contributions from new energy sources, such as solar and synthetics, prior to 1990, even if mandated by Congress.

Thus, within this framework of knowns and unknowns, a substantive issue that remains is, to what extent oil imports from OPEC sources can be curbed by increased domestic oil production and non-OPEC soft currency imports, and by reversal of the trend of substitution of oil for gas in the industrial market.

A New Marketing Strategy for Gas

A major change in marketing strategy for natural gas and a concurrent change in state and federal regulatory policies will be necessary to maintain and expand industrial and power plant sales. The image of gas as a scarce commodity was enhanced by the 1976-77 winter supply crisis and by the subsequent efforts of the Administration, Congress, and the regulatory community to restrict future gas use to high-priority categories by a variety of legislative and regulatory measures, including incremental pricing of new, higher-priced supplies for low-priority industrial purposes. Naturally, fuel oil filled the gap left by the unavailability or supply uncertainty of gas and the inability or unwillingness to convert to coal with all of its problems.

Although there is now growing understanding that the 1976-77 winter crisis was an aberration and growing belief that the decline in lower 48 discovery rates can be reversed by the provisions of the 1978 Natural Gas Policy Act incorporated in NEA, this positive view of future gas supply is far from universal. To be successful, a change in gas marketing strategy to seek major industrial loads again will require convincing proof to the regulatory agencies, Congress, the general public, and present and potential large-volume users of gas that:

1. More than adequate supplies exist, or are clearly in sight to serve all high-priority markets, especially new residential hook-ups, for the foreseeable future.
2. Substantial supplies in excess of high-priority needs are certain to be available for an extended period.
3. It is in the high-priority customers' economic interest to restore and preserve the high load factor characteristics of the gas supply system by continuing gas service to most industries and even selected power plant markets.
4. It is in the public interest to reverse the substitution of oil for natural gas in the industrial and power plant markets, and halt the substitution of electricity for gas in the residential/commercial markets.
5. The price levels needed to assure adequate supplies of gas for all its traditional markets, while high, are still competitive with all other environmentally acceptable options to meet growing energy needs.
6. It is in the self-interest of industrial and power plant users to buy gas from domestic and other secure sources rather than oil that is increasingly supplied from hard currency, foreign sources subject to *force majeure* interruption and cartel pricing policies, even if gas is priced at a premium; and failing that to demonstrate that it is in the public interest to mandate gas use by these users when available.

Needless to say, I believe that a strong case *can* be made with potential large-volume users regarding future reliability and relative price stability of gas service. However, the form value of pipeline-quality gas supplied from an integrated national supply and storage grid vis-à-vis oil, synthetic liquids, medium- and low-Btu gas, and coal in terms of operating economy and flexibility and environmental impact will have to be quantified more accurately. The power plant market offers a particularly attractive target of opportunity, since the electric utility industry is seriously considering various coal synthetics as a substitute for oil.

In devising a new gas marketing strategy, the drastic and continuing changes in the spectrum of supply sources and in the competitive structure of the energy market must be taken into account. The old strategy of achieving seasonal load equation was to sell surplus gas on an interruptible basis at a substantial discount over oil. This may no longer offer a viable long-term option. Even if rolled-in pricing for industrial gas sales could be extended to categories nominally ineligible under the recently enacted legislation because of the wide discretionary powers given to the Secretary of DOE and the Federal Energy Regulatory Commission (in some instances subject to Congressional consent), discounted sales on an interruptible basis may lose economic justification in view of the rapidly rising marginal cost of replacement and new sources of supply.

By 1985, if the gas industry is successful in expanding supply to as much as 25 trillion cubic feet, much of the gas will be from high-cost sources. The incremental price of new lower 48 gas, Alaskan gas, Canadian and Mexican gas and imported LNG in the appropriate market areas is likely to be $5 per thousand cubic feet in constant 1979 dollars. This is equivalent to oil of $30 per barrel, which may or may not be competitive with the constant dollar prices for industrial fuel oil prevailing at that time.

Therefore, this gas should be marketed as a firm or guaranteed seasonal commodity at the substantial premium over oil that industrial gas deserves and receives in Europe and Japan. The main competition for gas in the residential/small commercial market is electricity, which should comfortably allow a rolled-in delivered retail gas price of $5 per thousand cubic feet in constant dollars in 1985, even at a 3:1 efficiency advantage for electricity. The retail price of No. 2 fuel oil in the 1979-80 heating season passed $6/million Btu, so gas should have little problem to maintain its price advantage.

Thus, recapitulating, the major current issue relating to the role of gas in U.S. energy policy does not seem to be whether adequate supplies of pipeline-quality gas to stem oil imports can be made available, but how legislation and regulatory policy will affect the development and marketablity of these supplies.

The Role and Utilization of Coal in Increasing U.S. Energy Self-Sufficiency

It is apparent that implementation of a program to achieve a substantial degree of energy self-sufficiency for the United States by the year

2000 can depend only marginally on nuclear energy. This is due to the extremely long lead times for nuclear plant construction and commissioning, the excessive capital requirements of substituting nuclear-based electricity for direct uses of fossil fuels, and the exceptionally severe institutional problems exacerbated by the Three Mile Island accident, all of which inhibit efforts by utility managements to accelerate the rate of expansion of nuclear capacity. Maximizing the rate of development of indigenous oil and gas supplies can provide additional time but, ultimately, creation of a synthetic and biomass fuels industry will be essential to offset the scarcity and cost of imported oil.

Unfortunately, creation of a synthetic fuels industry based on the nation's abundant domestic coal and oil shale resources also requires substantial lead times and faces obstacles which, with the exception of the safety and waste disposal issues, do not seem much less severe than those impeding the growth of nuclear energy.

The alternative of direct coal substitution for oil and gas rather than via a gasification, liquefaction or solvent refining step has also made little headway because of the associated high costs of dealing with environmental and equipment conversion problems. As a result, coal use has remained demand limited.

The failure to achieve any substantial coal substitution so far and the deferral of optimistic goals for increased coal production have led to a reassessment of prospects for coal. It is now better understood that any realistic assessment of various options for achieving U.S. independence from oil imports must consider the total cost of substitution. This includes not only production, transportation, and distribution costs, but the entire downstream costs of changes in energy supply. Even with considerable incentives, the costs may turn out to be prohibitive. For example, omitting the cost of conversion of existing appliances, equipment, and facilities to alternative energy sources, the cost of creating new energy transportation and distribution systems, and the many hidden environmental costs of fuel substitution may lead to a completely false picture. A good case in point is the substitution of coal in the form of electricity for the direct use of oil and gas in many stationary heat energy appliances.

Coal Utilization in the Form of Synthetic Gas

Taking such matters into consideration leads to the conclusion that high-Btu pipeline-quality gas from coal is frequently a preferred method of substituting coal for oil and natural gas in such stationary uses because:

1. It can employ commercially-demonstrated technology at known and acceptable cost.
2. Good prospects exist for substantial improvements in existing technology on the basis of well-developed second generation technology.
3. It requires no costly and time-consuming conversion of natural gas-using equipment. This consideration is vital because natural gas is still the dominant energy source for homes, commercial installations and industry. Pipeline-quality gas can also be substituted easily in most oil-burning installations. In addition, high-Btu gas produced from coal can replace natural gas and, in some instances, oil, as a process feed where necessary.
4. The environmental impact of substituting coal for oil and natural gas by means of producing high-Btu gas from coal at the mine-mouth and delivering it through the existing underground pipeline system will be less than that of any other method of substitution.
5. Also, as noted before, high-Btu gas from coal can compete with conventional coal-generated electricity for heat energy requirements even if it is assumed that the end-use efficiency of electricity may be up to three times greater than that of gas.
6. It is not practical or economical for any but the larger industrial customers or base-load central power stations to substitute direct combustion of coal for the use of oil and natural gas because of the space and equipment needs, plus the high cost and frequent lack of technical feasibility of installing the necessary environmental control equipment. Also, where gas or oil must be used for process reasons, direct combustion of coal has no role to play.
7. Based on projections from pilot plant operations, the delivered cost of solid solvent-refined coal will be about the same as the delivered cost of high-Btu gas produced from coal in second generation plants, although high-Btu gas is cleaner and more versatile. Similarly, coal-derived liquid industrial fuels produced with technology now entering or completing the pilot plant stage, seem to offer no cost advantage over high-Btu gas from commercial, first generation coal technology.

Although, in general, it is as economical to use the high-Btu gas route to substitute coal for oil and natural gas as it is to use most other methods, there are some notable exceptions:

1. The direct combustion of coal with stack-gas clean-up or sulfur-removing fluidized-bed combustion systems is undoubtedly going to be cheaper in large new boiler installations where these pollution control methods are feasible.
2. Where solid solvent-refined coal can meet air quality standards in existing large coal-fired boilers, especially in central power stations, it may offer a viable alternative to other retrofit pollution control methods.
3. The use of low- or medium-Btu gas from coal in new combined-cycle power generation plants would undoubtedly be more economical than the use of high-Btu gas, and may become the preferred method of central power generation from coal in the future if certain technical obstacles are overcome. In fact, coal-based combined-cycle power generation may prove to be the toughest long-term competitor to gas-fueled options in premium energy markets.
4. Given a limited geographical area with a high concentration of heavy industry needing gas or other clean fuels, plus an opportunity for off-peak sales to electric utilities, the preferred method of substituting coal for oil and natural gas could be large, central, low- or medium-Btu gas installations.

However, in relatively small, dispersed industrial installations, low- or medium-Btu gas produced from coal will be at least as costly and have a greater environmental impact than high-Btu delivered by pipeline from central mine-mouth plants, because of the high cost of delivering coal, the poorer load factor, the poorer economy of scale, and the difficulty of providing adequate environmental controls for a multiplicity of small coal conversion plants.

Resource Base Considerations

Estimates of the remaining economically recoverable supplies of conventional natural gas and petroleum liquids in the United States have varied widely during the last few years, in part because of the differences in wellhead price expectations. Estimates of reserves and resources of coal and shale oil, on the other hand, have remained relatively stable.

On a Btu basis, coal represents about 80 percent of proved reserves or remaining recoverable resources, and the ratio of the latter to the former ranges from about 5:1 to 8:1.

The remaining recoverable supply of conventional natural gas from proved, inferred, and undiscovered reserves and resources is, by most recent estimates, somewhere between 600 and 1,200 trillion cubic feet or quads; 800 trillion cubic feet is a reasonable basis for planning purposes. This excludes potentially recoverable "unconventional" natural gas resources variously estimated at 200 to 3,300 trillion cubic feet or quads.

The remaining recoverable supply of conventional crude oil is on the order of 150 billion barrels or 900 quads, although it could be somewhat more if significantly enhanced oil recovery (EOR) can be realized economically. Current reports in the literature of oil production indicate that enhanced recovery is proving to be quite expensive. As an upper limit, on the assumption of 60 percent ultimate recovery, a value of about 370 billion barrels or more than 2000 quads could be derived from a 1972 National Petroleum Council estimate of 810 billion barrels of oil originally in place. This would compare most directly to the upper limit of the combined potentially remaining recoverable conventional *and* unconventional natural gas resources, which is substantially higher.

If the combined demand for all fossil fuels in the United States were to grow at about 2 percent per year, the total remaining recoverable fossil fuel resources (excluding unconventional natural gas) would last for another 110 years, leaving a 10-year forward reserve. If total fossil fuel resources could be stretched by the use of an equivalent amount of nonfossil energy forms and imports, they would last more than 140 years at a 2 percent annual growth rate. Even at a sustained 3 percent annual growth rate, there is more than enough lead time to develop the wide range of non-fossil energy sources, if coal and oil shale are fully utilized to replace the much less abundant gaseous and liquid hydrocarbon fuels.

Supply Elasticity

The estimates given above for conventional natural gas and crude oil resources relate to quantities judged to be economically recoverable without any rigorous specification of cost or assumed improvements in technology. Recent government studies (MOPPS) indicate a strong interrelationship of recoverable resource quantities and marginal cost, that is, a strong price elasticity of supply. If, for example, upper limits of $5 per thousand cubic feet and $30 per barrel are set for marginal costs of conventional gas and oil, lower 48 recoverable resources would be estimated as 600 trillion cubic feet or quads, and 75 billion barrels or

450 quads, respectively, on the basis of the "moderate" MOPPS supply curve projections, as modified from the original 1975 dollars to 1979 dollars.

In developing an overall energy policy, the degree of supply elasticity of our remaining hydrocarbon resources is a critical factor. With most synthetic fuels costing on the order of $4-7 per million Btu or $25-40 per barrel oil equivalent at the point of production (1979 dollars), it is obvious that the quantities of domestic crude oil and natural gas recoverable at equivalent wellhead prices becomes a major issue.

Views on the supply elasticity of conventional hydrocarbon resources vary widely. It is my opinion, on the basis of unpublished econometric studies, that the weighted average wellhead price of conventional lower 48 crude oil and natural gas (that is, the average on a Btu basis of all vintages, old and new, regulated and deregulated) would have to increase at a rate in excess of 10 percent per year *above* inflation in order to increase gross additions to reserves substantially above the current low levels.

Imposition of new tax disincentives may very well raise this already high threshold. Some experts believe that a positive supply response can be achieved with lesser incentives, such as those provided in the Natural Gas Policy Act. However, even average wellhead prices for conventional hydrocarbons as high as $5 per million Btu (in 1979 dollars) by 1985, which would very likely evoke an increase in gross additions to lower 48 conventional gas reserves from the current 10 quads annually to 20 quads (equivalent to 5 million barrels per day of oil imports), might turn out to be in the consumer interest, given most other options for secure and environmentally benign sources of energy supply.

The supply elasticity question becomes even more important when "unconventional" sources are included. According to the MOPPS study (as modified from the original 1975 dollar base to 1979 dollars) these resources up to a wellhead price of $5 per million Btu would add 650 trillion cubic feet to potential supply, for a total of 1,250 trillion cubic feet, or 60 to 65 years at current rates of consumption.

These unconventional sources of natural gas not included in standard resource estimates comprise 200 trillion cubic feet of natural gas in tight formations in the Rocky Mountains and natural gas contained in the Devonian shales of Appalachia. Perhaps an additional 450 trillion cubic feet of methane may be recoverable from coal seams and from the geopressured brines of the Gulf Coast area within this price limit. At wellhead prices higher than $5 per thousand cubic feet, the resource base for naturally-occurring methane is believed to increase exponentially and, in the opinion of some experts, astronomically.

Need for Synthetic Fuels

Obviously, overall energy policy should include the full development of these naturally-occurring hydrocarbon resources at a cost competitive with alternatives. However, this is not an argument for deferring the development of synthetic sources of gas.

The reason is that most supplemental sources of gas, be they natural or synthetic in origin, are cheaper and often more readily available than their liquid fuel equivalents. This would include tertiary recovery of crude oil vis-à-vis unconventional natural gas, commercial Lurgi gasification vis-à-vis Fischer-Tropsch synthesis of liquids, synthetic liquid fuels from coal and oil shale via hydrogenation and retorting processes vis-à-vis high-Btu gas from second generation gasification processes, etc.

Therefore, it would seem to be in the national interest to fill as many stationary heat energy needs as possible with natural and synthetic gaseous fuels, even including many central power generation needs. It should be stressed again that the nation's oil import problem was caused to a considerable extent by the reverse of this substitution process, namely, the replacement of natural gas with imported oil. If these stationary uses of oil were eliminated, there would be a lot more domestic oil available to meet transportation fuel and certain petrochemical needs, the only markets where liquid fuels are really essential. In this connection, it is reassuring to note that something close to existing technology could provide supplemental supplies of motor gasoline from coal at a pump price quite competitive with what consumers pay in most of the world.

The Role of Imported LNG

For both the United States and the rest of the world, conventional natural gas reserves are about equal in magnitude to crude oil reserves. However, the production rates of these two fossil fuel resources differ considerably. Based on 1976 data compiled by J.D. Parent of the Institute of Gas Technology, world crude oil reserves are being produced at a rate of 3.5 to 3.9 percent per year, whereas world gas reserves are being produced at a rate of 2 to 2.3 percent per year. Excluding North America, remaining world proved gas reserves are being produced at only 1.2 to 1.4 percent per year. This indicates the potential for a significant expansion of gas reserves production to provide LNG to world markets.

Imported LNG could make a very significant contribution to U.S. gas supply. Projections by the Institute of Gas Technology and Chase Manhattan Bank indicate LNG imports could reach as high as 2.4 trillion cubic feet per year by 1985—55 percent of a world trade of 4.4 trillion cubic feet per year. The technical and safety problems have been solved, U.S. companies have been eager to implement LNG projects, and world energy requirements are supplying the momentum for further expansion. The real roadblocks are regulatory and governmental. In the past, the Federal Power Commission's slowness in approving applications for projects has resulted in project delays or termination. Incremental pricing has also been a roadblock, although recently the Commission has recognized that imported LNG must be sold on a rolled-in basis in order to ensure the financial viability of such projects.

Apprehension has been expressed in government circles about the national security implications of increased reliance on LNG from some of the same suppliers that provide the less secure supplies of OPEC oil. A good bit of this apprehension can be dispelled when the economics and logistics of producing and delivering such oil and LNG are compared.

First, in contrast with crude oil production, LNG requires massive capital investments on the part of the producing nation. These funds can only be obtained from external sources on assurance of uninterrupted debt service. In spite of this larger producer investment, the balance of trade is more favorable than for oil imports because of the large U.S. ownership of relatively costly LNG tankers. Second, the designs of liquefaction and terminal facilities and tanks are closely integrated for any given project to minimize costs, so that diversion of LNG shipments is much more difficult than of crude oil and its products, and the cost of providing for extended storage of LNG would be prohibitive. Therefore, barring some major international catastrophe, LNG supply interruptions are much less likely than oil supply interruptions.

Plan for Maximizing Gas Supply

The provisions of the 1978 Natural Gas Policy Act, combined with an increased availability of federal leases, should temporarily arrest the decline in lower 48 natural gas production. Under these circumstances, a production level of 19 trillion cubic feet appears feasible in 1985. Hoped-for Alaskan supply could provide 0.9 trillion cubic feet. Pipeline imports from Canada should continue to increase somewhat and Mexican imports should become a significant supply source by 1985, for a total of 1.5 trillion cubic feet. Based on announced projects, it should also be possible to realize 1.5 trillion cubic feet per year of LNG imports by 1985. Natural gas from unconventional sources could start making a contribution to supply by 1985—0.5 trillion cubic feet in addition to the amount now included in conventional production. At least two coal gasification pioneer and demonstration plants are likely to start production by that time, but total supply from these sources is not expected to be significant. Although substitute natural gas from light petroleum distillates and natural gas liquids (SNG) could provide one quad per year by 1985, it would be risky to place any major reliance on this source for the longer term. All in all, a total supply of 24-25 trillion cubic feet is achievable in 1985 if an all-out effort were made.

In order to achieve a total pipeline-quality gas supply of at least 30 trillion cubic feet by the year 2000, gas supply from all sources must be maximized. Production of conventional gas from the lower 48 states is likely to decline from the 19 quads per year level in the post-1985 period due to resource base limitations. However, it may be possible to keep the rate of decline at a sufficiently low level to realize 15 trillion cubic feet from this source in 2000. Alaskan gas and pipeline and LNG imports could provide at least 4 trillion cubic feet. Timely development of natural gas supplies from unconventional sources could add up to 8 trillion cubic feet in 2000 at competitive prices according to a recent study for DOE by Lewin and Associates. SNG produced from light hydrocarbons largely for peaking purposes can be assumed still to provide 0.2 trillion cubic feet in 2000. An annual synthetic gas-from-coal capacity of about 3 trillion cubic feet of pipeline gas equivalent would then be needed in 2000 to bring total gas supply to 30 trillion cubic feet. However, between 1985 and 2000, it still appears feasible to build as many as 40 high-Btu gas plants of 250 million cubic feet daily capacity each (40,000 barrels per day oil equivalent), plus 20 medium- and low-Btu gas plants of 150 billion Btu daily capacity each. Aggregate investment cost in 1979 dollars would be about $75 billion, or $5 billion per year, to provide a total capacity of 4.3 quads of syngas in 2000. This investment rate is small compared to historical annual electric power plant construction budgets.

To achieve these objectives, a coordinated effort by industry and all branches of government will be required. In addition to the creation of a viable synthetic fuels industry based on coal and oil shale, with major emphasis on production of SNG and transport fuels, the other essential elements of this effort are:

- A regulatory climate that permits gas to compete fairly in all energy markets on the basis of relative form value, security of supply, benefit to the U.S. economy, contribution to monetary stability, environmental impact, and capital requirements;
- No erosion of the incentives for maximizing exploration, drilling, and production of conventional gas supplies in the lower 48 states and Alaska;
- Lowering of the barriers to accelerated leasing of federal lands, particularly federal offshore areas and federal land in Alaska;
- Acceleration of projects to transport Alaskan gas to the lower 48 states;
- Resolution of legal, institutional, regulatory, and policy issues restricting the importation of LNG;
- A greatly accelerated program of research, development, demonstration, and commercialization of gas production from the unconventional gas resources in the geopressured brines of the Gulf Coast, the eastern Devonian shales, the Western tight sands, and coal seams.

All of these steps are essential if the goal of at least 30 trillion cubic feet of gas is to be achieved by the year 2000. For the longer range, the expansion of the synthetic fuels industry and the development of inexhaustible sources based on biomass and solar energy are essential to the continued availability of gas as a major energy source.

Even if we can maximize production of conventional and unconventional natural gas supplies, the urgent need will remain to gasify coal, our most abundant fossil fuel resource, to satisfy the economic demand for gaseous fuels.

With parallel programs for maximum feasible development of our domestic oil supplies, synthetic liquid fuels production, nuclear power generation capacity, direct use of coal, and the uses of the many forms of solar energy, timely reduction of oil imports to manageable levels should be possible.

No set of policy initiatives would be more beneficial to the U.S. public and the economic and social progress and political stability of the world as a whole.

The views expressed by Dr. Linden in this paper are solely his and do not necessarily represent those of Gas Research Institute or its advisory bodies.

Discussion

From the floor: Will you comment further on the so-called "gas-bubble"—how large it might be and how long it might last?

Linden: The "gas bubble," which is temporary excess deliverability of natural gas created by regulatory restriction of gas markets, has gone down in size from two quads to six-tenths of a quad. (I will use annual quads here as the equivalent of annual trillion cubic feet of natural gas.) Two quads used in one year would replace the equivalent of one million barrels of oil a day. The source of the bubble is the oversupply that built up in the intrastate market—largely in Oklahoma, Louisiana, and Texas—during the time we had a dual system of regulation, i.e., a free natural gas market in intrastate commerce and a controlled market in interstate commerce which has now disappeared thanks to the Natural Gas Policy Act. This excess intrastate supply was exacerbated on a national scale by federal and state regulatory policies that restricted industrial use and new hookups of gas for residential and commercial use, and legislation in the Fuel Use Act that required that natural gas had to be phased out of all power plant use and most industrial use by 1990.

In other words, because of a misreading of the 1976-77 winter gas supply crisis as a basic gas shortage rather than the result of unpredictably severe weather, the response of many governmental bodies was to restrict markets for gas. When you restrict markets, you eventually get an over-supply.

The numbers are very simple. The residential/commercial market, which has the highest priority, uses only 7½ quads of the 20 quads of gas that are now available annually. And by 1990, residential/commercial demand will be only 9 quads out of whatever the 1990 number is for total gas supply. That would leave a lot of gas for industrial use under any foreseeable supply scenario. Allowing only high priority gas sales and keeping even this market relatively static by restricting hookups, etc., depreciation of the huge transmission and distribution system would have to be applied to smaller and smaller quantities of gas to the detriment of the residential and commercial consumer, which would further accelerate pressures to legislate and regulate gas out of the energy market.

So, the gas bubble was a creature of state and federal regulation and legislation, and it will last maybe another year or year-and-a-half. The basic issue is, is the American Gas Association, is James Schlesinger, is the Department of Energy right about U.S. ability to maintain gas supplies at the 20 quad level? There are lots of models around. My own model—which, since I am a chemical engineer, is not very credible—tells me that the answer is very price-related. My model says that the Natural Gas Policy Act will provide, from now until 1985 when most gas is to be decontrolled, only a 10 quad finding rate of conventional gas in the lower 48 states. That is at less than $3.00 per million Btu (1979 dollars) regulated wellhead price. However, I believe that at $3.50 per million Btu, the lower 48 states finding rate would be 15 quads of gas, and a $5.00 price would bring that to 20 quads. Five dollars in constant 1979 dollars in 1985 for gas will be a bargain compared to any other alternative of equivalent form value. Moreover, this 20 quad finding rate, which will result in a 20 quad sustainable production rate, can be supplemented with Alaskan gas, imports, etc., to get the nation up to the 25 or so quads that will be needed in 1985-1990.

The gas bubble is transitory. Discovery will go back on the downhill track unless there are price incentives for conventional natural gas. The relationship between finding rate of oil and gas, and the deflated average wellhead price of oil and gas, and the cost of drilling wells is very well-defined. The average wellhead price of old and new oil and gas has to be increased by about 10 percent a year above inflation to keep finding as much as has been found in the past.

From the floor: There has been quite a bit of discussion in industry on super-projects; the gas industry is probably looking at eight or ten such projects of about a billion dollars each. If you had to prioritize four or five main ones that policy should strive to see completed, which would they be?

Linden: I am going to give what you may consider an evasive answer. In the hierarchy of energy supplies, all these projects are competitive with alternatives of equivalent form value. The equivalent of pipeline gas from coal is refined synthetic liquids from coal and oil shale; medium-Btu gas from coal is equivalent to making synthetic solids and liquid fuels for power generation and large industrial uses. In that frame of reference, gas is competitive, and the proposed projects should proceed. Imports of natural gas are similar: gas from Canada and Mexico, which are soft currency, secure Western Hemisphere sources, is vastly superior to oil imports from overseas. So all of these projects should go ahead. And the Alaskan gas, although it is going to be expensive—in 1979 dollars, perhaps $6.00 a million Btu—is useful as part of the total mix, and this increment of supply should be developed. That these projects are not going ahead at a reasonable rate hinges on the fact that there is no reward for risk-taking in the national interest. We have no system for this. That is why I proposed a scheme that puts everybody under a legislative mandate to market at least 5 percent (or whatever the appropriate figure might be) of pipeline gas and liquid fuels from non-conventional or synthetic sources by 1990. To do this on an ad hoc basis with loan guarantees, tax credits, purchases or price guarantees, etc., is a very slow process, as we have seen.

From the floor: Do you have a feel for the dollars per Btu that it will cost to get at the geopressurized zone gas? And how much is there?

Linden: It has been estimated to be equivalent to 3,000 to 50,000 quads; I do not believe the economically recoverable amount is anywhere near that high. A set of studies has been made on very meager evidence, and the indication is that up to a 1975 dollar wellhead price of $4.00 (which is like $5.00 now) per million Btu, something like 150 to 200 quads could be recovered, which would nearly double the current proved reserves of natural gas. I emphasize that this is a "ballpark" number, not something definitive. My organization, jointly with the Department of Energy, is now in a drilling and testing program costing on the order of $35 million a year, to come up with some numbers that will validate the potential supply and economics of this and other unconventional gas sources.

From the floor: What is your feeling about the solar option?

Linden: It has to be pushed, and pushed very hard, to keep as a reserve option, because if the nuclear option fails, which it could in the light of much current evidence, and if the fossil fuel option (which could see us through another 100 years at 2-3 percent annual demand growth) is limited because of the atmospheric CO_2 build-up problem, then what do we have? The question that must be answered is, must America then regress to a *low* technology solar economy a la pre-1750 with mules and horses and only firewood and farm wastes for fuel and fertilizer, or can it go to a *high* technology solar option that can indeed produce aluminum and steel and plastics and transport fuel, and all the things that are needed to maintain social and economic progress—all with solar energy? It is a very important question, and I think as in so many things, the populace, the public feeling—uninformed as it is—the gut feeling of the people is more correct than that of the experts. I do not know how it would be accomplished, but do not sell solar short.

From the floor: Leaving aside the Alaskan gas pipeline, do you think the existing gas utility industry can provide capital to develop the coal gasification and other unconventional sources, or do you see other industry supplying that?

Linden: The evidence so far is not too positive. One reason that electrification has been so successful is that the electric industry is vertically integrated and has been able to finance its high rate of growth without outside help. It maintains a spinning reserve of 20-30 percent in spite of many problems. The gas industry has not done that. It is not vertically integrated, and depends for much of its supply on others. It has a history of curtailments that goes back to World War I and World War II. So there is a question. If the gas industry can prove that it can deliver clean and useful Btu's more cheaply from secure sources, it is in the consumer and national interest to help develop new sources of supply. Certainly the Natural Gas Policy Act has helped; it increased conventional gas prices, decontrolled deep natural gas produced from below 15,000 feet, and decontrolled most unconventional natural gas sources. The question is, will this be enough?

From the floor: What is the reality of the prospects for liquefied natural gas in view of the action of the Energy Regulatory Commission in killing a couple projects?

Linden: This has been a very difficult issue. The evidence is that DOE, including the FERC and the ERA, is now more flexible. They have changed their mind over the last two years on the value of more LNG imports, and I think that energy needs and economic realities will prevail. But I think there will not be too many new terminals. There already are two on the Eastern seaboard that can be expanded—remote terminals that provide the operating safety needed. There will be a similar one on the Pacific Coast and maybe one on the Gulf Coast. I think as much as two quads of LNG could be imported annually between now and the year 2000, in spite of this terminal shortage and the prices that will have to be paid. And it will be a useful source of energy.

Chairman: What you have been telling us—and I will simplify it to saying we should continue a dependence on natural gas at about 25 percent of total energy supply—I find very convincing, and I am wondering what keeps it from happening? What stands in the way of getting total cooperation toward what you are recommending?

Linden: One problem that stands in the way is the pricing and marketing of industrial gas. What is needed is a system that will give large-volume users reliability of supply so gas can command the premium of as much as 30 percent at the burner tip over oil which it commands in Europe, for example, and in Japan, where new power stations are being built in high-population areas with costly liquefied natural gas as fuel. Gas must be made the fuel of first resort rather than last resort.

In the residential and commercial markets, it is necessary to get over new hookup restrictions, and that is underway now. But the industrial market has to be segregated in a way to give reliability of supply, which means a lot more storage for the low load factor residential/commercial markets, so industrial gas can be sold on a relatively firm basis at a premium, instead of increasing our dependence on imported fuel oil subsidized by entitlement payments. That is underway too.

Bruce Murray

The Case for Solar-Electric

I will argue the case for the production of electricity directly from sunlight, and I begin by acknowledging that, in the narrow sense, electricity produced in this way generally is not "economic." Indeed, if one looks at the costs of capital equipment for solar-electric compared to nuclear and coal-fired systems, it is hard to prove solar-electric systems will be economic before the end of the century.

However, when one takes into account all the social consequences, especially the safety and health hazards of nuclear energy and the health and environmental effects of coal, the actual costs of using coal and nuclear systems have consistently been much larger than acknowledged by the industry. Thus, I feel the actual economic gap between solar electric systems (which exhibit fewer health, environmental and safety problems), and nuclear and coal-fired systems will narrow considerably. In some regions of the country, the economic gap is likely to disappear altogether.

Of course, no one can confidently *predict* what will happen other than that the United States is going to suffer a very large readjustment in its sources of both electricity and heat, and that prices will increase substantially over time. Just how plentiful energy will be in the future

---

At the California Institute of Technology where he has been associated as researcher and educator for close to two decades, ***Bruce Murray*** *is Director of the Jet Propulsion Laboratory, one of the country's most important contractors in the field of solar energy about which he spoke. A geologist with three degrees from the Massachusetts Institute of Technology, he has received NASA's Distinguished Public Service Award for his work in space research, which included participation in five of the Mariner experiments in planetary exploration.*

remains to be seen. What the mix will be is a variable. Given such unknowns, our concern should be to understand what benefits and liabilities the different alternatives offer, and how they may best fit into the infrastructure of the country.

As implemented by the Department of Energy, the country's national energy strategy should be to provide appropriate subsidy and infrastructure incentives to develop a diverse mixture of new energy alternatives, so that the society that makes the choices five, ten, fifteen, and twenty years from now has something good from which to choose that will meet its real needs.

Heat and Electricity from Sunlight

Solar energy will certainly be used for heating. It is already so used, and can be deployed much more widely—starting with simple solar water heating and moving into space heating. But what about the use of solar energy to produce electricity in significant quantities by the end of this century?

The production of electricity from sunlight is more complex. Many people are not optimistic that new technology options such as photovoltaic cells or solar-thermal electric systems (that concentrate sunlight to produce heat in order to run an engine to produce electricity) will prove to be sufficiently useful and competitive in the future to generate significant amounts of electricity. I believe they will, but primarily in a decentralized and modularized way, not generally as part of very large-scale systems used for central power generation. Central solar electric systems have some of the problems faced by coal-fired or nuclear plants. They require very high initial capital investments and create a substantial land use question as well. I think there will be some such systems in time, but I do not believe they will be the dominant system for the production of electricity from sunlight, at least not within the coming two or three decades.

Satellite Solar Power Systems

Similarly looking to the future, some of my space colleagues have developed the even more challenging concept of using satellites involving huge solar-electric systems in orbit. Such systems would have some technical advantages. They could collect sunlight that otherwise would not hit the earth, and then dissipate the heat, mainly in space instead of on the earth. They could be the source of a nearly continuous supply of energy, providing base load capacity. They would most likely be tied into central electrical distribution stations through "rectenna farms" near cities.

There are, however, uncertain environmental effects and many substantial technical questions which, by comparison with other new energy alternatives, are really unknown. In any case, satellite solar power systems would be the most costly new energy sources conceivable in the near- to mid-future, because neither the launch vehicle capability nor the space manufacturing processes required have been developed. Hence, for the remainder of the century, satellite solar power exhibits the same kinds of problems as do conventional nuclear and coal plants. The proper near-term strategy for satellite solar power systems is to develop the space-based tools and capabilities that may enable this option to be explored as a serious alternative after the turn of the century.

Small-Scale Applications

For the nearer term, I think development efforts for the production of electricity from sunlight should be targeted on relatively small-scale applications such as shopping centers, factories, and schools. The solar-electric systems would be co-located with the consuming entities themselves, on roofs or on land, under control of the user. Then the question of land use becomes a trade-off made by a single user concerning his own land, and the very difficult and subjective issue of using one interest group's land for someone else's benefit is avoided.

Solar energy is not well suited to providing base load for the obvious reason that during nighttime, no electricity is produced. So it will first be most attractive in places that have a high summer peak load. In Southern California, for example, there is some natural matching of solar availability with heavy summer daylight demand for air-conditioning.

These small-scale solar-electric units would be tied into the electric utility distribution system. Electricity would have to flow both ways down the utility wire. The same distribution system that delivers central generation electric power would have to accept power from decentralized solar generation unit in times when the system was not being used at a maximum by the institutions in question, for example, on weekends. That novel two-way relationship raises questions of utility pricing structure. Eventually, this and other infrastructure questions will need to be addressed so they do not become major barriers to the introduction of solar electricity.

Advantages of Solar-Electric Systems

Technically the kind of system I am talking about would be of low modular cost. Furthermore, the cost of going into the business of producing solar-electric modules would, by many energy project standards, be very low.

Relatively Small Cost

For example, the capital cost to build a plant that each year could manufacture a total of 250 megawatts* encapsulated photovoltaic panels (for sale to others for installation and tying into the utility) is estimated to be about 50 million dollars. Thus, when the time comes that it is perceived to be economic to go into the business of manufacturing solar panels, there are many corporations with the financial strength to do so. In the interim, the government must continue to subsidize research and development to provide ever more efficient manufacturing technologies until natural market forces take over.

I want to make it clear that the absolute amount of capital needed to go into the solar-electric system production business is relatively small. Similarly, relatively small investments are needed to buy photovoltaic or thermal-electric systems in modular fashion.

Local Freedom of Choice

Another aspect of solar-derived electricity gains importance if the United States continues to evolve into a fuel-allocated society, namely, the local freedom of choice permitted. Fuel allocation has clearly surfaced. Gas lines are a manifestation of allocation of gasoline to dealers by producers. At the consumer level, highest priority for the available allocation simply goes to those who wish to stand in the line the longest. If the United States goes into a fuel-allocated society in one fashion or another, whether it be coal, gas, or oil, the use of supplementary solar-electric is an additional advantage because its fuel cannot be subject to allocation.

Other questions one might ask are, "How much sunlight is available?" or, "If one did all the necessary things, what fraction of the energy needs of a particular area could really be taken care of by solar energy?"

---

*Peak power capacity.

Potential in Particular Regions

Let me give an example that is not a practical one but constitutes a useful mental calibration. If, in the case of the San Fernando Valley, a suburb of the Los Angeles megalopolis, all the southward-facing roofs on factories, businesses, and homes were equipped to collect sunlight, and the sunlight were used to produce either electricity or heat, and there were some way to provide adequate storage (which we are not going to talk about), almost all the current stationary energy consumption in that particular area would be provided. Thus, for a suburban area with the climate, level of commercial activity, industrial development, and population density of the San Fernando Valley, solar energy has the potential to be a very significant part of the total energy mix.

In fact, I do not envision that degree of "solarization," and I do not want to overemphasize the example. Yet, it does suggest that in particular regions the useful energy content of sunlight coming down on existing roof structures is substantial, and could provide a significant part of the energy mix. On the other hand, in places such as downtown Manhattan there would not be enough roof tops for collectors. There, in order to use solar electricity significantly, one would have to commit large areas of land (or sea) surface elsewhere.

Cost in Relation to Coal and Nuclear Energy

"How much will solar electricity cost?" Based upon the estimated cost of components and of producing the various silicon materials or other devices that are used, solar-electricity is expected to cost in 1986 about a factor of two higher than electricity from coal-fired plants. Thus, the apparent electrical generation costs perceived by the customer, one way or another, is about a factor of two higher. This is the principal reason solar energy—and especially solar-electric—is not thought to be a major energy alternative before the end of the century in the United States. Why would people want to pay twice as high an electric rate in one form or another?

At the Jet Propulsion Laboratory we believe that solar-electric cost will be reduced. By 1986 we foresee photovoltaic systems reaching cost parity in both residential and intermediate lead center applications. By 1990, cost parity in utility applications is achievable. Furthermore, I believe that the probable costs for the nuclear and coal alternatives are going to increase much faster than inflation, principally because society is finding undesirable side effects from them and is requiring that many expensive steps be taken. The current cost of installed coal and nuclear power is much greater than had been forecast, and that pattern is likely to continue throughout the period of time we are considering here. As a consequence, the perceived cost advantage of coal and nuclear, with respect to solar-electric, is deceptive. There are very few side effects about modular, decentralized solar-electric technology that are likely to be attached in the same way.

Excess social costs of coal and nuclear power are tacked on in the form of licensing regulations and antipollution control measures. Let me give you some brief examples. The installed cost of nuclear-electric generation increased about a factor of three between 1965 and 1975. Initially, this was primarily due to technical and production bottlenecks. But since 1970 it has been almost entirely due to the actions of interveners over safety, health, and environmental issues. The resulting delays have so run up costs that the probable front-end capital expense of a new nuclear plant may make it "uneconomic." Orders for new plants have come to a complete standstill. In fact, even before the Three Mile Island accident, there had been more cancellations than new orders. That pattern could change in the future, of course; but it is not going to change unless something fundamental happens to change that front-end capital problem, i.e., the delays and uncertainty in licensing.

Consequences of Commitment to Coal

New coal-fired plants constitite the key to the country's energy future, according to the Department of Energy. Originally, United States production and utilization of coal was to be increased by a factor of four by the end of this century. A more modest factor of 2½ is now being mentioned. In fact, the production and utilization of coal in the United States have not increased significantly in the last decade, even during the last five years when such an increase has been the goal of national policy.

The reason it has not happened is its difficulty. There are formidable technical problems of extracting more coal, and all kinds of associated social costs as well. For example, there is the problem of transportation: the present rail system is barely adequate to transport even existing coal production. There will be a need to go to slurry pipelines and other capital-intensive innovations. And those have environmental and other social consequences that will be showing up as additional costs and delays.

Another point: coal utilization produces a very large amount of sulfur dioxide and other pollutants. It is important to remember that a vast amount of solid material is involved. In the "good" situation, only 1 percent of coal is sulfur. In a bad situation, 5 percent is sulfur. And sulfur forms some noxious compounds, many with serious handling problems.

Effect of Social Concerns on Cost

So the consequences of doubling the use of coal in the United States are profound and, in my view, have not been thought through. In any case, the same effect of social concerns on the cost of new coal-fired electrical generation plants are turning up that have been seen on the cost of nuclear energy. Between 1965 and 1975, the installed cost of coal plants about doubled. During this period, the percentage of cost due to pollution control in one form or another rose from 10 percent to 42 percent. It is estimated that by 1985, two-thirds of the installed cost of coal plants will be for some form of pollution control. Coal, therefore, is following the same pattern as nuclear power production, only a little bit later in time. The government has made a significant verbal commitment to coal as *the* solution to our problems. The commitment is similar to that of the old Atomic Energy Commission to the nuclear power industry. I do not see much evidence that a great deal has been learned from our own energy history. Organized political action groups are convinced that nuclear power is morally wrong. They have not yet said coal is quite so pernicious, but the time seems close.

On the other side of the argument, people in government and the industrial sector are saying, "How can anyone be so stupid? Coal is the cheapest way to go. It's in the best interest of the nation to use it." With that degree of polarization in a free country, what actually results is stagnation and relentless escalation of the real cost. And I am afraid that is what will happen with coal also if the United States tries to double its use on an accelerated basis without a period of expensive and difficult preparation of the entire system.

Whether my gloomy prediction as to the increasing costs of nuclear and coal power (due to the burgeoning of social costs through regulation, taxation, and delay) comes true in a broad form or only in a more moderate form, cost increases certainly will continue. On the other hand, the production of electricity from sunlight has fewer potential social cost drivers. This is especially the case if the user of solar energy makes his own decision on how to use it. He is the one who suffers the environmental consequences, if any, of its use. Accordingly, it is very important for the country aggressively to develop the technology for this option, although not necessarily commit to its widespread use until later.

The Decentralized Solar-Electric Option

Developing the decentralized solar-electric option means not just subsidizing the development of manufacturing technology but also seriously trying to work on barriers to introduction that arise from the infrastructure of the country itself. For example, current utility rates are designed to promote expanded use of centrally generated electricity. They are not designed to optimize the mix of central and decentralized sources. That can be done, but it is a question of what incentives the utilities are given. The present incentive structure tends to inhibit the introduction of decentralized sources.

Another barrier to the use of solar electricity is the relatively high front-end cost to customers who want to purchase their own solar electric equipment. Of course, annual operating cost is quite low. Somehow, the question of lifecycle cost has to be taken into account in a practical way for the small- and medium-sized user.

Development of a Solar-Electric Market

One approach that appeals to me is suggested by the way gas water-heating and other gas systems were introduced into an initially unprepared market. The gas utilities marketed and sold gas-burning appliances, supplied maintenance, even leased equipment, to help get the new energy system installed. It is easy to imagine utilities providing financing for solar-electric systems or even retaining ownership in the systems and selling an "energy source." Thus, it could be possible for consumers on a large scale to choose solar-electric or other elements of solar energy quickly and effectively as part of their energy mix.

Unfortunately for solar energy, the country is torn by the kinds of polarization that are exhibited with nuclear. Some of the people who are the strongest advocates for solar energy also have the deepest suspicions of utilities and other large institutions in this country. They favor aggressive development and early use of solar energy but at the same time emphasize individual actions with minimum involvement of large corporations, particularly the regulated monopolies and the multinational energy companies.

This attitude is superficial and can even be counter-productive. The only way solar energy will be installed on a large scale in the United States is by either use of the present utilities or erection of a government bureaucracy to perform the same functions, in which case the usual difficulties accompanying an extension of government bureaucracy can be expected. Similarly, large-scale, low-cost production of solar-electric modules must be based on modern automated manufacturing processes; it simply cannot be done by a cottage industry.

Pricing Considerations

The other important change required to accelerate the introduction of solar energy in the United States is that the price of energy to the consumer must, in fact, be set at its replacement cost. President Carter has taken some very important steps toward that end—first, with the deregulation of natural gas, and second, the phased decontrol of oil. But he is not getting any help from Congress. It is a measure of the confusion in the country as to what our national purposes and needs are. I believe that President Carter and the Department of Energy deserve a very strong endorsement for persevering on what is the most important thing—an increase in the retail price. Energy price increases are not justified, in my mind, to stimulate the finding of more oil or natural gas. All that does is accelerate the depletion of our limited hydrocarbon reserves. Price increases *are* justified, first, to induce conservation and second, to make other energy technologies and alternatives more attractive economically. They can be fully attractive only when the price of the existing energy sources reaches their replacement cost.

So my conclusion is that sunlight very well may be a practical and significant option in the direct production of electricity in the United States—before the year 2000 as well as afterward. Technically, it will be available for that purpose. Economically, it may well become practical because of the continued problems in making increased production of coal and nuclear electricity socially acceptable. The enhancement of the availability of solar-electricity depends to a large extent on how well the infrastructure barriers are recognized and overcome in a manner analogous to the technological barriers. It is probably in the arena of social and political events that the solar energy scenario will succeed or fail.

E.C. Davis of the Jet Propulsion Laboratory provided many helpful suggestions concerning this paper.

Discussion

From the floor: Is there not a fundamental or serious flaw with solar energy because you must store it when things are dark?

Murray: The most important initial use of solar energy for electricity, in my view, is as a supplementary source tied into a central grid system. The grid is the storage. When there is excess solar energy being produced by the distributed collectors, it is fed back into the grid and it reduces the requirement for intermediate or base load at that point. Conversely during nighttime or bad weather when the distributed solar energy sources are not producing electricity, the base and intermediate loads have to pick it up. So solar comes in as a supplement; there is no storage required in that case.

The next step is to put in fuel cells and partial storage of various kinds, and one gets into a rather complex situation of various kinds of storage at various levels versus using the central grid system entirely. But for the application I described—San Fernando Valley is a good example—there is no need for any storage. That is an economic choice versus other kinds of trade-offs.

I repeat, I am not talking about solar as a replacement for central generation of electricity. I am talking about solar as a *supplement* that will reduce the need for new central generation capacities and new intermediate load in the future. *That* is where the issue will be. We are not going to decommission our existing central generation plants. The issue is, how many more coal and nuclear plants will be introduced, and it is in that context that decentralized solar-electric generation tied into the grid becomes a very interesting competitor.

From the floor: You did not say anything about what every Department of Energy study in renewable energy came up with in electric power shortage, and that is other than biomass—wind energy. Why is more not being done about this?

Murray: I was talking only about *direct* forms of solar energy. Wind and biomass are indirect forms. Of course, they also come in a decentralized way. The answer to the question is that much of the same logic I gave for photvoltaics applies to wind and biomass. Just as in the case of direct production of electricity from sunlight, very strong regional factors enter in. There are a sun belt and a frost belt. I anticipate that California and Arizona will use solar-electric generation as a significant factor in their local and regional solutions. Wind energy may well become important in other parts of the country that have relatively predictable and high velocity winds. As for biomass, in the long haul it is a very important solution, an area in which real scientific and technological breakthroughs are possible, especially through artificial gene manipulation of the microbial products involved in processing the biomass. This is probably one of the real breakthrough technologies we can look forward to in the future. How long it will be, no-one knows, but certainly it deserves a very high priority.

From the floor: I have a question on solar heating. I am on the board of a very large cooperative bank in Washington, and we had a competent professional engineering firm tell us that for our building, solar was not economic within a factor of five of current costs. I am puzzled because I thought you said there is now technology that will produce solar-electricity at about twice the cost of electricity from coal.

Murray: I do not know the specifics in the Washington picture, but it is true that costs are regionally variable and that the significance of solar heating at the present time, especially supplementary to another heating system, varies greatly. At the present time, solar-assisted gas water-heating is at least an even bet in California. Solar-assisted space heating is more marginal, and both are very strongly influenced by taxation and regulatory policies and the utility pricing structure. What I am saying is, right now they are competitive in Southern California and are being put in. They are not being put in as well as possible, but that has to do with the institutional means of doing it, not with the economics. I cannot comment about the situation in Washington, but there are places right now where solar-assisted heating is, in fact, quite an economic proposition and is being used. In the matter of the cost comparison, I was talking about electrical generation, and I said the projection for 1985 is that the difference in the cost of electricity produced from sunlight from that produced by coal, is a factor of about two in the favor of coal. And I was commenting that one has to recognize that the real cost of coal at that point may be a good deal higher than the projections because there are many more social costs still to be added in.

From the floor: What are the social costs that solar may see that have been omitted in the modelling?

Murray: It is difficult to find very many, frankly, except the land use, and that is why the example I chose was not one where the whole Mojave Desert is used for the benefit of the San Fernando Valley. There is no problem of vulnerability of transmission, for one thing. That is how those land-use questions will probably get worked. The idea that somehow city dwellers will fill up the rest of the country with solar collectors to solve their high energy use is probably very naive. No-one is advocating that, nor basing any thought on that.

From the floor: According to studies I have read, there is a plan in effect that is to develop the production capacity of photovoltaic units on a large scale at a cost of 50 cents per peak watt in 1975 dollars. In your opinion, how can that be achieved by 1985?

Murray: Let me make two statements. First of all, if the growth goes in a reasonable way, it *can* be done. We at JPL believe that—and JPL is in charge of that plan to which you refer—and all the evidence we have is that we are on track. But in order for it to happen by 1985 there must be decisions made by many private sector groups, both to manufacture solar-electric power and to use it. There is no way of knowing what will happen. A national strategic decision will have to be made by DOE and other people as to the point at which it is reasonable to create new incentives or to reduce the disincentives that exist. That is not what I am arguing. I am not saying it is going to happen. I am saying that the country should go as far down the road as seems reasonable to develop the option. A lot of factors, like whether Mexican gas and oil become available, will influence at what rate the productive capacity should build up. But the model we are using and are still on track on (because it does not require large-scale private investments yet) is a logical and self-consistent model.

From the floor: Why are people not paying more attention to geothermal?

Murray: The answer, I think, is a political one: because it benefits only a limited number of places. If there were geothermal deposits all across the country it might be looked at with greater government interest.

Chairman: Then why is there not a major geothermal program in California?

Murray: The State of California has, in fact, given very high priority to geothermal, and has also tried to put maximum pressure on DOE to get federal funds applied for that purpose, so I would say geothermal has been given adequate priority within the resources California has.

From the floor: The point should be made that geothermal is everywhere in the world. A hole drilled down into the earth will get to a temperature at which steam can be made. That does not mean in the geyser area alone. It is anyplace, right under your feet.

Murray: On the other hand, the ability to extract that to useful form in service depends upon the fluids that are involved and the porosity and a lot of other factors as well as the depth. In fact, any reasonable aggregation of what is the most easily available geothermal and therefore the most economic has very strong regional variations. In the Gulf Coast of Mexico one would have to drill about a hundred thousand feet into the crust of the earth, and I do not think that is an economic procedure.

From the floor: You said that society has consciously to provide the incentive structure that will bring together the decentralized and centralized approaches, and so I suppose that "incentive" system means disincentives as well as sanctions. My question is, Who gets what?

Murray: This is a good question on which to end because it picks up on the statement that is the most important part of my message, which is that the real outcome will be influenced by the incentives, both positive and negative, that decision-makers at both the delivery end and the consumer end will have offered them. I tried to emphasize that the present system is tuned to a *central* generation and delivery system using very high density energy sources like nuclear and coal and oil, and that if we are going to move to a source that is *decentralized*—whether it be wind or biomass or direct conversion from sunlight—the incentive structure will have to be revised. I do not know whether it will happen, but I do know it is a requirement. And I am asserting that if the incentive structure does get revised, then very substantial progress will be made very quickly. This, therefore, is a societal question much rather than a technical one.

J. Clayburn La Force

The Political Economy of the Energy Crisis

I see our nation's *real* energy problem as consisting of the following: The demand for energy is growing and the cost of developing more of it is rising.

Therefore, prices of energy sources, particularly petroleum and natural gas, will probably rise more rapidly than prices of other goods over at least the next two or three decades. Increasing energy prices are a problem; they induce costly changes in consumption patterns and production processes. These are inescapable and necessary changes.

Prices in a Free Economy

I want to emphasize that the resource problem—whether it be the problem of a less available resource, or the problem of having that resource available at higher prices through the operation of a successful cartel—is something that our economy can, and will, work out and adjust to, *if allowed to*. And this is a very big "if," because the history of oil and natural gas in the last fifty years in this country is a nightmare of complex and irrational regulations. The regulations are in direct contradiction to allowing the economy to make the necessary adjustments quickly and in an orderly manner.

J. Clayburn La Force *is Dean of the Graduate School of Management, University of California, Los Angeles, and former Chairman of the University's Department of Economics. He is a board member of the Hoover Institution on War, Revolution, and Peace and of the National Bureau for Economic Research, and is on the Research Council of the Center for Strategic and International Studies, Georgetown University. Dean La Force is Trustee and Chairman of the Foundation for Research in Economics and Education.*

You may have noticed that I did not say that our energy problem is that we are about to run out of fossil fuels. Some people are skeptical of the claim that we do not have a shortage of energy that will severely limit energy consumption and force a drastic reduction in our living standards over the next fifty years. The truth is, for at least 500 years experts have been predicting energy doom in the near future, usually in ten to twenty years. Many are making the same prediction today. They are as wrong now as other seers were in the past.

The solution to our problem is simple. Here it is. Prices in an unregulated market economy will (if allowed) efficiently direct the appropriate adjustment of consumption and production to a world of rising energy prices. The government does not need to subsidize home insulation through tax credits. It does not need to tell automakers to improve fuel economy. It does not need to tell industry to switch from oil and gas to more abundant fuels such as coal. It does not need to tell homeowners to adjust their thermostats. Individuals can and will make all of these adjustments with maximum efficiency and without having to be coerced by the bureaucracy. Sermonizing by the White House is also redundant. Even more unnecessary—I would say ludicrous—is Attorney General Bell's recent call for Americans to develop a "national sense of shame and practice self-denial to solve the energy crisis."

Past Government Regulation

I shall return to my solution later. But first I want to deal with government regulation of petroleum. I want to demonstrate that bad public policy over the last four decades has gotten us into our current mess.

Control of Domestic Production

One of the interesting things about the petroleum market in the United States is that at nearly all times since 1935 it has been manipulated. We see two distinct eras; the first from 1935 to 1971, and the second from 1971 onward. Regulation of crude began during the Great Depression. State regulatory commissions estimated the demand for crude and used this estimate to regulate the output of every well. In the late 1950s the federal government added restrictions on imports of foreign crude. The purpose of these quotas was to get higher prices for domestic producers. They were successful. In addition to higher prices for crude, we got an energy lobby, a depletion allowance, and a lot of richer oilmen.

Import Quotas

The oil import controls helped domestic oil producers at the expense of both energy consumers and the entire economy. First, higher domestic petroleum prices discouraged United States consumption. Everyone was "too" careful with energy. Those who refer to the '50s and '60s as a period of wasteful and excessive use of petroleum are wrong. In fact, during this period the artificially high prices of crude resulted in factories using less of this kind of energy, in houses receiving more insulation, and so forth.

Paradoxically, as the artificially higher prices discouraged domestic consumption of crude oil, the import quotas excessively encouraged domestic output. This happened because restricted imports drove up domestic prices. This in turn made it profitable for producers to produce crude from fields in the United States that they otherwise would not have exploited.

Clearly, the quotas on imports of foreign crude and the depletion allowance's tax subsidy artificially stimulated oil production in this country. Indeed, if our domestic producers had had that used-up supply at the time of the Arab embargo, they could have offered OPEC quite a bit more competition.

Recent Government Regulation

In the early '70s the second era of regulation began. The political tide turned against crude oil producers. A new combination of special interests succeeded in changing policy. During this era, the goal of nearly every policy-maker has been to encourage domestic production, discourage consumption, and promote the cutback of imports of expensive foreign petroleum. Their actions, however, actually have achieved the *exact opposite*. Let me illustrate.

Price Controls

The administration of President Nixon gave the crude producers their first big reversal in 1971 by deciding to phase out the quotas. At the same time, depletion allowances were ended. Both policy changes made domestic crude production less profitable. Even more damaging to producers of crude, however, were price controls. President Nixon initiated economy-wide controls in August 1971. And while price controls on most goods and services were phased out during 1972 and 1973, they remained on petroleum.

The combined effects of price controls and the end of favorable regulatory treatment of the oil industry have resulted in sharply reduced domestic production of crude oil. During most of the 1970s, production has been falling at about 4 percent per year. The discouragement of domestic production, in fact, has made us especially vulnerable to any sharp reduction in the world supply of crude and has been a boon to OPEC.

On the other hand, at the time of the Arab embargo, price controls on petroleum products artificially restrained the supply of gasoline. This was because refiners could not afford to pay the dramatically higher price for crude and still send as much gasoline to the market as customers demanded at the exceptionally low controlled price. People lined up for hours to try to get a share of the artificially restricted supplies. Plants shut down and workers lost their jobs.

I want to emphasize that lines at gas stations and industrial shutdowns in response to fuel shortages were the result of federal price controls. They were not caused by the OPEC price increase. I should mention here that the lines now forming at gas stations are a consequence of government's interfering with the market price of gasoline. In the absence of price controls, the embargo and subsequent increases in prices of crude would restrict the amount demanded and ration the available supply for those who valued it the highest. Such market-determined prices would be *higher* than controlled prices. But these higher prices would signal the increased scarcity of petroleum. We consumers might complain about higher prices when we buy gasoline or industrial petroleum products. Yet such prices do exactly what is required—they tell us to reduce our consumption of petroleum products because they are now more scarce.

The obvious effects of the energy shortage were caused by price controls on refined products. But these controls were effective for only a few months. By mid-1974, market prices were generally below controlled levels and the shortages therefore disappeared. Prices of domestic crude remained controlled, however. The effects of the controls on crude have been less visible than those on refined products, but they have been far greater. They have, of course, discouraged domestic production. But they have had other less obvious, yet important, effects.

Emergency Petroleum Allocation Act

Let us investigate one curious consequence of the controls on crude. Please focus your attention on refiners of crude. The 1973 Emergency

Petroleum Allocation Act froze the existing buyer-seller relationships between refiners and domestic producers of crude. So a refiner who was buying a relatively large portion of his crude at about $5.00 per barrel from suppliers subject to price controls, suddenly would have lower average costs of production than would a refiner who was purchasing most of his crude from OPEC countries.

The fixing of crude oil prices, allied with the freezing of the crude oil buyer-seller relationship, caused very substantial differences in profits among refiners. A gigantic amount of wealth that would have gone to domestic producers of crude after the OPEC price increased went instead to some lucky refiners.

Entitlements Program

Enter then the Entitlements Program—a bizarre gimmick for altering this outcome. Adopted by the Federal Energy Administration in 1974, the Entitlements Program aims to remove the profit disparity among refiners. The Federal Energy Administration issues monthly entitlements to controlled crude oil to each refiner.

In short, the effects were that those refiners originally benefiting by the controlled price and the frozen arrangement between buyers and sellers of crude were forced to compensate the other refiners not favored by these regulations. If the FEA sets the number of entitlements right, it will equalize the *average* crude oil price paid by all refiners and wipe out the original impact on refiners' profits.

The Entitlements Program also subsidizes the importation of foreign crude oil. In short, refiners see the cost of imported oil as being less than the OPEC price. Consequently, refiners buy more foreign oil and increase their production of refined products—just as if they had received a subsidy from the United States government. Foreign sellers of crude oil benefit by selling more oil. Consumers of refined products benefit from the expansion of product supplies and resulting lower prices.

By the time Jimmy Carter was elected, the Entitlements Program was paying for about one-fifth of the then $13 per barrel OPEC price. This was worth about $7 billion a year to foreign sellers of crude. The greater availability of refined products and lower prices added about $7 billion per year to the wealth of users of petroleum products. Meanwhile, price controls on domestic crude oil producers deprived them of approximately $14 billion per year!

From a government preaching energy independence came massive discouragements to domestic energy production and huge subsidies to the international oil cartel. From a government preaching demand restraint and conservation—a government that would tell us how to heat our homes and what kind of cars to drive—came subsidized prices for gasoline, heating oil, and other products made from petroleum. Regulation of the petroleum industry was and is a *cause of, not a cure for,* the energy crisis in this country.

National Energy Plans of Carter Administration

After such a dreary record, few economists expected any quick moves toward an efficient, rational policy by the Carter administration. And the Carter proposals were indeed the same old irrational policies that had been in effect for at least forty years. The few parts of his first energy package that got through Congress—such as the natural gas regulations—exacerbated the situation. Fortunately, much of what he wanted did not become law.

The President's national energy plan called for continuation of crude oil price controls, with some new wrinkles. President Carter would have had the 14 billion dollars per year, which I said was transferred away from petroleum producers as a result of price controls, controlled directly by a federal tax on crude oil. Under this scenario the administration would then have had 14 billion dollars to hand out—supposedly to energy users in the form of income tax rebates.

Within the past several weeks, President Carter has unveiled his second attempt to develop an energy policy. Of course, it is too early to tell what he will end up with. Basically, he has decided (actually events have forced him) to remove price controls on domestic crude. The price of all domestic crude will be allowed to rise to the level prevailing on the international market. A very important benefit will be the withering away of the Entitlements Program—that inane system of regulation that hugely subsidizes OPEC as well as the output of refined petroleum products.

Windfall Profits Tax

A second major benefit might be the stimulation of domestic production. But the President's strident call for an excess profits tax places that potential benefit in jeopardy. Producers of domestic crude will gain substantial revenue by being able to sell their outputs at the market price. However, a combination of the federal corporate income tax, state and local taxes, and a windfall profits tax could take as much as two-thirds of that increased revenue. The costly process of searching for and bringing to market new petroleum depends upon profits.

In addition, the President's alarmingly virulent rhetoric in recent weeks should have unsettled even the most intrepid of investors in this industry. I mean such statements as: "The windfall profits tax is the only thing that stands between the oil companies and a huge bonanza of unearned, unnecessary and unjustified profits." Or this one: "I will not allow this painful but necessary step to become the excuse for a massive rip-off of the American people by the American oil companies."

Perhaps the windfall profits tax is a necessary political facade. It is, however, an unambiguous revenue move. And Congress and the President will have many billions more to spend in politically determined ways. That tax, along with the harsh demagoguery, does little to improve the outlook for increased production of petroleum in the United States.

Conclusion

When I started, I said that the underlying source of concern over energy is that prices of energy will rise faster than those of other goods and services. In the short term, these price increases reflect the impact of OPEC. Over the long term, they will reflect the increasing scarcity of, and higher costs of extracting, fossil fuels.

I also have argued that the increasing costs of energy should not be a source of panic or despair. We in the United States are not going to run out of energy, huddle together in the darkness for warmth, and feel our world grind to a halt. If market incentives are allowed to operate, the economy can adjust as supply and consumption patterns respond to higher energy prices.

However, I have stressed the crucial role of federal energy policy in determining how well we make those adjustments to high and rising energy costs. From my analysis of the recent history of policy and my conception of the political realities and struggles that lay behind energy policy, I cannot help but be pessimistic about the future, despite the President's decision to decontrol the price of crude.

An energy crisis is a debacle created by government policy. We remain on a path toward increasingly regulated energy markets. Please remember that we still have the Department of Energy. It has more than 20,000 employees, a $12 billion annual budget, and is about as likely to be disbanded as Congress itself. We still have, and shall continue to have, controls and regulations.

Such policies for dealing with the scarcity of energy, I firmly believe, are poor substitutes for the ubiquitous and subtle incentives in an unfettered, free economy. Not only do controls and regulations distort necessary adjustments to higher energy costs and impair the overall health of the economy; ultimately, they restrict individual liberty. They are in stark opposition to the tenets of a free society.

The author wishes to thank Joseph Kalt for assistance in preparing an earlier version of this paper.

Discussion

From the floor: Will you comment on the proposal included in the President's package to tax the difference between domestic and OPEC prices?

La Force: You mean the plan where, if the OPEC price is 16, say, and domestic crude is decontrolled so it goes to 16 also, and then OPEC goes to 20, two dollars of that difference goes to the government? That would be an extremely bad tax, in my opinion. I would argue strenuously against it. I see no reason for it. It is obviously a revenue package for the government. It certainly does nothing to provide incentives for the companies. I see no economic rationale for it.

From the floor: I would be interested in your response, as a free market economist, to the program outlined by John Foster for creation of a synthetic fuels capacity, which—as I understand it—would be a kind of lifeboat arrangement to provide a cap on future energy prices. It would involve something on the order of a \$300 billion investment to produce some considerable quantity of synthetic fuels by the year 1995. At the present rate of spending by the financial community and the lack of a proven running industry, there seems to be no way to accomplish this except with major government involvement. How would you look on that kind of proposition?

La Force: If it had to be done, I prefer the alternative suggested by Charles Hitch—where rather than direct government production of these plants there would be bids by private industry to go into the business. I have a deep conviction that the market is more efficient than individuals can be in making choices and decisions. I would like to see price incentives existing for people to make the investments. Maybe they would not be made, but I am wary of government getting into this business.

Chairman: On the subject of government influence on energy by way of controls, sponsorship and funding, versus free enterprise and investment, how would you, as a free market economist and an educator, answer if your students were to ask, isn't there some important area of energy other than weapon systems that requires government to do the investing and take the risk? Do you really think it all can be done by free enterprise? All of it?

La Force: Let me put it this way: What operational principles would we use to determine where the government would be involved? I have no idea. I think it would result in people lobbying for projects, and we may or may not come out with something that would be a winner. But if companies and individuals are allowed to work on their own, with their own funds and motivated by the possibility of profit, thousands of probes would be going on in a search for alternative energy sources. This does not mean the government has no place. But it has to allow this to occur.

Henry Linden (from the floor): Is there not a specific type of enterprise that seems to be related—communications and other networks where it has been government nearly exclusively for development?—the railroad system, opening of the West, air transportation (which is largely a creature of the development of the jet engine and its application), the franchising of utilities—electric, gas, oil—isn't there a simile there? None of these things would have come about without massive government intervention.

La Force: But hasn't this been with proven technology already? Was the government involved with the creation of the technology or did it already exist?

Linden: It was involved in the *commercialization* of the technologies—heavily. Certainly the railroads, certainly the jet engine, and certainly the interstate highway system. I just am puzzled how energy, which is a major component of the utilities, can be exempted.

Chairman: Looking at the present and the future, and putting aside what theoretically might be possible, in actuality there is no chance of our being a totally free enterprise system so far as energy is concerned; there is no question but that government has to be in the act. And because of that, free enterprise is not going to risk massive investment in synthetic fuels with the certainty—as investors might feel—of government control, of unknown environmental requirements, of changes in rules, and so on. Therefore, the only way a project of this size is going to happen is as a hybrid. Whether you think it is right, or the best way, is that not true?

La Force: Quite possibly it is the *only* way. Maybe that is the way it has to happen. What I have been saying is that the policy we have had so far with the petroleum industry has been very bad; it has had just the opposite results from those claimed. And I was limiting my remarks to the petroleum industry. I was calling for restraint, the removal of controls in an effort to allow the market to operate. If the market is not allowed to operate, obviously industry is not going to produce the synthetic fuels. They won't drill the oil, either.

From the floor: I should like to extend Henry Linden's comments. There *has* been a role for government subsidy of technology where public benefit is great and the costs more than the private sector could manage—that is how we got not only jet aircraft but also microelectronics. So what one looks for is some mechanism that allows combining the market system with the usefulness of government as an aggregator of market power. What do you think of the suggestion that I heard from a number of people, that government structure a project in synthetics in such a way that it is outrageously profitable, and then allow private companies to bid for the right to do the project? Obviously the prices they offer will reduce the profitability until they compete the project down to some appropriate rate of return, considering the risks and so on. Is that idea consistent with your notion about market projects?

La Force: It is certainly far more consistent than having the government actually plan the investment and build the plants. It would be far preferable, so far as I am concerned.

From the floor: Another interpretation of the synthetic fuels dilemma is that the reason private industry will not take it up is that it is basically not a good idea and is not going to pay off—that synthetic fuels would have such low profitability that they are not worth investing in for either government or the private sector.

La Force: I have no idea whether it would be a good or a poor investment. One reason private industry might not want to get involved is because there would be a period of price controls and regulation that would *make* it unprofitable. But whether there is a possibility of eventual profitability, I do not know.

John E. Swearingen

Tomorrow's Energy: The Future Is Now

Our topic is the future of energy in the United States, and when we speak of our energy future, we must also by definition speak of our future as a nation. Plentiful and timely supplies of energy built our democratic civilization as we have come to know it here in this country. Without adequate supplies, our future cannot realize the promise of our past; and today the future is nearly upon us.

Albert Einstein once had this to say: "I never think of the future. It comes soon enough." And indeed it does, as many of us have discovered in our daily lives. But to relate Dr. Einstein's maxim to personal considerations is one thing, and to apply it to the formulation of public policy is quite another. Yet, I feel that is precisely the case with our national energy policy. Our government allows short-term political considerations to override long-term economic solutions to our energy problems; as a result, these problems worsen. In the process, the quick-fix political solutions achieve a permanence of their own and create a whole new generation of problems that follow them. Thus our energy future to a very considerable degree is ransomed to our political present.

In an era of luxury and excess such as the decades of the '50s and the '60s, this may have seemed of no great concern, but as the grasshopper discovered, you cannot sing and dance all day and not eventually pay

---

John E. Swearingen *is Chairman of the Board of Standard Oil Company (Indiana). A Phi Beta Kappa graduate of the University of South Carolina, he is a member of the National Academy of Engineering, a fellow of the American Institute of Chemical Engineers, and a trustee of Carnegie Mellon and DePauw Universities. He has headed the American Petroleum Institute and the National Petroleum Council.*

the price. Now, as we approach the decade of the '80s, the bills are rolling in. As a well-known football coach, George Allen, used to put it, the future is now, and the time has nearly passed for us to wake up and face it.

In the area of energy (and America *is* energy) we must face up to some grim realities. While we have been "singing and dancing," the situation has continued to deteriorate. Finally, however, the federal government seems to have begun to realize that the time has come to pay the piper.

This realization is reflected in some recent statements issuing from the White House. I must say that I take issue with the tone of those statements and the rhetoric in which they are couched. I understand the political considerations involved—1980 is a Presidential election year, and the oil industry provides a convenient target for tub thumping. But political posturing will do nothing to solve our energy problems. Even so, when the political considerations are factored out (and that is indeed a formidable challenge of its own), many of President Carter's recent pronouncements demonstrate at least an awareness of the dimensions of the nation's problem.

In his latest address to the nation, for instance, the President said, "Our energy problem is very serious, and it is getting worse." Those of us who work in the energy industry have been saying this for years, but nonetheless it is somewhat heartening to realize that the government in Washington may finally be awakening to see things as they are.

In the same address, President Carter had this to say: "Our national strength is dangerously dependent on a thin line of oil tankers stretching halfway around the world, originating in the Middle East and around the Persian Gulf, one of the most unstable regions in the world." That is a lesson all Americans should have learned more than a half-decade ago when the Yom Kippur war broke out in the Middle East and much of our supply of crude oil was embargoed. The lesson did not take. At the time of the embargo the United States was importing approximately one-third of its oil. Today the import figure is nearly 50 percent, and it took a cold water shower provided by the Iranian crisis to awaken the nation to the profound dangers of that growing dependence.

Over-Regulation Discourages Production

What is the cause of U.S. overdependence on foreign sources? There is only one answer. For much of this decade, a complex and irrational web of governmental controls, laws, and regulations has held domestic prices of oil and gas well below world market levels; and because prices have been held at artificially low levels, the funds to finance expanded domestic exploration and production have not been available. So, inadvertently or otherwise, Washington devised a system that discouraged production, encouraged consumption, and relied on imports to fill the gap. The Administration's first energy plan, announced some two years ago, did absolutely nothing to rectify this situation.

Today there is some reason for optimism. The President has apparently faced up to the seriousness of our energy dilemma. He has come to grips with the supply side of the supply-and-demand equation by announcing his intention to decontrol domestic crude oil prices. A program to phase out crude oil price controls is long overdue, and allowing domestic prices to reach world market levels will accelerate the search for domestic supplies. In the world of economic realities, price must reflect cost, and decontrol will allow that.

Having said that, let me make a caveat: as everyone knows, the President is also calling for Congress to impose a windfall profits tax on oil industry revenues. None of us can know what Congress will eventually devise, but what the President has proposed is to tax away half the revenues generated by higher oil prices. In my opinion this is both unnecessary and counter-productive. It is unnecessary because under current law about 60 percent of those additional revenues will go to state and federal governments in the form of taxes and royalties. It is counter-productive because when added to normal income taxes it would increase government's take to more than 75 cents out of every additional dollar of revenue, thus leaving the industry without adequate wherewithal to intensify its search for and production of new domestic supplies. Finally, why should the oil business be singled out for special taxes? The price of houses today is twice what it was ten years ago. Is there any clamor for taxing away windfall profits on houses? Of course there is not. And the reason? It is because there are more homeowners who vote than there are oil producers. This is a political matter pure and simple.

Contradictory Policies

There is a pattern here that has been seen time and time again in the energy industry. Washington issues a statement of policy which other announced policies then cancel out. Two years ago the Administration made conversion from oil and gas to coal one of the centerpieces of its energy program. Yet, at the same time, it enthusiastically enforces rules and regulations that make the mining, transportation, and burning of coal virtually impossible. It allows industry to spend millions of dollars in applying for permits and building nuclear power plants, and it also allows opponents of nuclear power to prevent those plants from operating, by endlessly litigating the same old questions in court. (Let me add here, incidentally, that I am not sanguine about the future of nuclear energy as a viable energy alternative. The opponents of nuclear energy have always been highly articulate and well-organized, as those who live in California certainly know. And now the Three Mile Island incident has lent additional force to what was already a highly emotional issue.)

The same situation pertains across the energy spectrum, be it the construction of refineries, or the search for new supplies of oil and gas. Refinery construction has ground to a halt because of various governmental strictures, and this despite the Department of Energy's insistence that we need new capacity. Even as the President calls for more domestic exploration and development, he is also locking up, as wilderness, millions of acres of public land that might be productive of oil and gas.

My purpose here is not to inveigh against the federal government, but I must insist that the energy future of this country will remain bleak if the government cannot get its energy objectives and directives organized, and it continues to demonstrate its inability to do so. The President proposes to encourage domestic production by decontrolling crude oil prices. He simultaneously calls for a tax that would defeat that very purpose. I am against windfall profits, and I am against them whether they accrue to industry or to the federal government. That is not the case here; and all the enflamed rhetoric and exaggerated figures coming out of Washington mouths cannot convince me that it is. It may be a windfall for government, but it certainly is not one for industry.

The Profit "Plow Back"

It is unfortunate that the President seems so profoundly misinformed about the role that profits play in industry. Profits are used for the very thing that President Carter says he hopes to achieve by decontrolling crude oil prices, namely, to find and produce more domestic oil and gas. For the past several years, my own company has been reinvesting in our business about twice the amount that we realized in earnings, and the same is true of much of the industry. Apparently, part of the President's problem is that he does not trust the industry to invest whatever additional profits it receives in the stepped-up search for new supplies in the development of alternative sources. No doubt that distrust is partly instinctive, but it is also partly political.

Here, perhaps, the record is illuminating. According to a recent Chase Manhattan Bank survey, 27 oil companies last year invested some 29.5 billion dollars in new exploration and capital investments. Of that total, only 2.8 billion was spent on non-energy areas; the rest, nearly 27 billion dollars, more than 90 percent, was plowed back into the energy business. And over a ten-year period the records show that 97 percent was reinvested in energy projects. That, I submit, should reassure the most angry populist critic; and that is the course the industry can be expected to follow if it is allowed to use the revenues that would come from decontrol of crude oil prices. And if our nation is to have any energy future at all, those revenues are more necessary than before.

What will those revenues be used for? Today, nearly three-quarters of America's energy supply comes from oil and gas, and that figure will not alter appreciably over the next ten years. Therefore, production from these conventional sources must be increased. It *can* be increased, but only if people are willing to pay the price. In the decade ahead, the industry will be drilling deeper, opening new frontiers, developing new sources of supply, and seeking to increase the recovery of oil from known fields through the application of new technologies. All these efforts will be costly, but given adequate revenues, I am convinced this industry can successfully pursue them as solutions to the near-term problem.

For the longer term, in the mid-1980s on into the 21st century, several possibilities exist for expanding energy supplies. One of these is the development of crude oil from oil shale; another is to produce oil from coal. It can be done; we in the oil industry know how to do it. We may not have the best and most economic process that will ever be developed to accomplish this, but we could start out today and build a plant that would operate satisfactorily to recover oil from shale or to convert coal to oil and gas.

When talking about these sources of supply for this country, one must keep in focus the magnitude of the task involved. It would cost about one-and-a-half billion dollars to develop a capacity of some 70 thousand barrels of oil per day from coal or from shale. If the object is to eliminate our present imports of foreign oil, we are talking about replacing 8 to 10 million barrels a day. That means a program costing some 200 billion dollars, one that would require eight to ten years for completion. This is an endeavor that would dwarf such great national efforts as the Manhattan Project in terms of time and money expended.

Furthermore, the money required can come from only one place: it must come from the users of energy, whether they pay the bill as consumers or as taxpayers. These are examples of the kinds of alternate energy sources that can be made available over the balance of this century if people are willing to pay the price. That price is high. But it must be considered as a form of insurance policy against the unthinkable alternative that U.S. domestic security and U.S. foreign policy will be completely at the mercy of the whims of foreign potentates.

Alternatives to Petroleum Economy

Further down the road there will be other alternative sources to supply America's energy needs, and among the sources most frequently mentioned are wind and tides and geothermal energy and solar energy; and of these, of course, solar power is the most intriguing. Solar and other forms of exotic power cannot be realistically viewed as workable alternatives until well into the 21st century, and to rely on them as a panacea is to take an ostrich-like view of mid-future supply problems. When one has his head in the sand, he is extremely vulnerable in other areas.

It should also be remembered, that while some of the best things in life may be free, solar energy and wind energy are not. It is necessary first to build and install the equipment to capture it. The cost of this energy will be several times that of fossil fuels.

I have no doubt whatever that alternatives to a petroleum-based economy do exist, as costly as they may be, but they are not available today and they will not be available in any significant quantity for the next two decades. Therefore, for the last two decades of this century, in my opinion the United States must go with what it has, and it has plenty. The total hydrocarbon base in this country is sufficient to carry the nation well into the next century. All that is required is a willingness to pay a reasonable price for its development. As the space program of the 1960s demonstrated, Americans can do almost anything they are willing to pay for. It is possible to increase the recovery of oil from known reservoirs, to produce oil from shale, to liquefy or gasify coal. It is possible to explore and develop the deep waters of the continental shelf and to exploit the methane in deep brines along the Gulf Coast entrapped in coal and shale beds.

(In this conference I have nowhere heard mentioned that the problem with gas entrapped in the deep brines in the Gulf Coast is, to get at the gas, the water must be lifted. The cost is in lifting the water. And then, what is to be done with the water when it gets to the surface? If it must be pumped back underground, that is an additional cost. The methane in the deep brines along the Gulf Coast, in my view, is still quite some way down the road as far as making any substantive addition to our supplies.)

In short, the oil industry can provide ample supplies of energy here at home, if the price of those supplies is allowed to go high enough to cover the cost of producing them and to allow a reasonable profit in addition. I think here of the price of gasoline. You, I, all of us cannot have tomorrow's energy at yesterday's prices and here lies the political stumbling block. For years, gasoline has been one of our nation's great bargains, and consumers have become accustomed to what seemed to be permanent bargain prices. Throughout the rest of the industrial world, nations such as West Germany, Japan, Italy, and France have seen the price of gasoline considerably higher than $1.50 a gallon. In France, for instance, the figure today is $2.50 for a gallon of leaded regular gasoline, and as yet I know of no discernible public outrage. I do not believe there would be a public outrage in this country if the people were told the truth about the supply situation by their elected leaders. The problem is one of economics and not of physical limitation, and I have no doubt that if the American people understood that they could have an unlimited supply of gasoline at higher prices, they would opt for those higher prices rather than do without.

(Incidentally, if the government really wants to reduce the retail price of gasoline, why does it not remove its taxes? I am not familiar with the situation here in California, but in Cook County, Illinois where I live, direct taxes—state, federal, county, regional transit authority—add better than 20 cents to the cost of every gallon of gas that is sold. If those direct taxes were to be removed, along with the numerous indirect taxes affecting gasoline prices, the price at the pump would be substantially less than it is today. But, as yet, I have heard no proposals along these lines issuing from Washington, and I must say I do not expect to hear any.)

"Insurance Policy" for Energy Security

Therefore, the only answer to our supply problem is higher prices. There will, of course, be those who argue that higher prices for petroleum products will add significantly to the growing rate of inflation. That, I submit, is a distinctly myopic point of view. Consider, for instance, the projected inflationary impact of crude oil price decontrol carried out over a three-year period. According to the best estimates available—and these are estimates not only by industry economists, but by the President's Council of Economic Advisors and the Department of Energy—the effect on inflation is measured by the consumer price index to be less than one percent over the three year phase-out period. In my opinion this is a small price to pay for a measure of energy security.

Furthermore, when the additional funds available from decontrol are invested by the industry in stepped-up exploration, the ultimate effect will be to lessen the nation's dependence on imports. That automatically strengthens the dollar, and a stronger dollar will have a deflationary impact on prices. So, in the long run, decontrol is the only sensible answer.

In the end, the economic aspects of America's energy problems are solvable, but the political obstacles will remain formidable as long as substantial numbers of our citizens believe they can have more energy without paying for it. Perhaps, as the new rhetoric issuing from Washington suggests, that illusion may finally have been shattered.

The shattering of an illusion can be painful in the short run. In the long run, the effect often tends to be beneficial. If the loss of illusion leads to public policies that recognize that the energy of tomorrow will cost more than the energy of yesterday, our nation and our allies will be the ultimate beneficiaries.

In terms of tomorrow's energy, the future is now.

Discussion

Chairman: You mentioned 200 billion dollars as the figure required for creation of a synthetic fuels industry. In the absolute, that sounds like a huge amount of money. As a fraction of the GNP over a period of ten years, it is not a large percentage; it is but a fraction of the defense budget, and energy supply is critical to U.S. security. Do you really believe it is possible to create a synthetic fuels industry? And do you see any fundamental limitations, like getting that much capital through private control?

Swearingen: There is no problem in undertaking to do this job with known technology, but in my view there is no way such a program could be begun in this country today and brought through to a successful conclusion in a period as short as five years.

After all, what is [an industrial] plant? It is a collection of furnaces and heaters and pumps and compressors and heat exchangers and things of that kind. There is a limited ability in this country to manufacture such items; there are not enough skilled workmen to do this in a crash period. But apart from that, during wartime, people [are willing] to make any sacrifices necessary in order to see that an essential program goes through. (I can recall giving up the spare tire on the automobile in the '40s to support the war effort.) But there is no consensus in this country today that the sacrifices necessary to put a synthetic fuels program in place will be accepted by the American people. And until there is, I do not believe that it could be accomplished in a short period of time.

The second question has to do with financing, which I think could be accomplished if the price level were allowed to go high enough to support the output of the plan. There are several ways this could be done. The government could provide loans for construction. The government could give tax credits against other income for this construction. The government could contract to buy the output at a given price. The government could enact tariffs or other duties in order to see that the products would be marketable at a price that would repay the investment that was made. These are only four of the ways that this could be accomplished. The money could be made available in one way or another to build this industry, if the American public decided it wished to do so. And in this connection, the best service that the President of the United States and many of our representatives in Congress could do would be to try to explain to the American people the seriousness of the problem we face.

Chairman: A number of those in the audience are concerned about the possibility of the energy picture in this country becoming even more grim, perhaps because at least one of our speakers, Dr. Bueche, suggested that probably things will have to get worse before they get better. With regard to the image that petroleum companies have with the public, and the way the petroleum companies' profits are used politically, one way in which things can get worse is that the oil companies might even be nationalized. Do you expect that?

Swearingen: This question finally boils down to whether [Congressmen] think the people already in the business can do a better job of finding conventional oil and gas, or developing new supplies of coal and shale than the federal government can. What I have observed leads me to believe that I do not want any postoffice mentality or railroad regulation mentality running the oil business.

Chairman: One person asks, "If the government would get totally out of the way, could free enterprise really be able by itself to resolve the energy situation?"

Swearingen: It probably could, but I believe the timing would be more rapid if the government recognized the problems that face the country and assisted with a program that eventually will have to be undertaken. After all, national defense is a function of the federal government and not of private industry. Consider what it would cost to establish in the Indian Ocean a counterpart of the Sixth Fleet that now is in the Mediterranean. If we Americans are going to depend on Middle Eastern oil supplies for the next generation, we will have to insure their availability. I am not a shipbuilder, and I do not have any precise figures, but I do know that aircraft carriers now cost three to four billion dollars, Trident submarines cost one to two billion dollars, and if it were necessary to build a base to service a fleet, my estimate is we would be spending somewhere between 25 and 50 billion dollars to protect our supply line. And the question I raise is whether it is not to our advantage to spend that money to develop alternatives to buying from the Arabs on a long-term basis.

Chairman: The point of view is put forward by some that there is evidence, and most geologists would agree, that there is no longer much U.S. oil [still in the ground], so, would it not be wise for petroleum companies to show a different kind of leadership, and instead of investing so much in looking for something that is presumably not there in adequate quantities anyway, to plow the money into a crash program to develop solar energy, for example?

Swearingen: The premise of your question is false. I do not accept the assertion that there is a limit to the U.S. oil industry's ability to find new oil and gas resources in the United States in the next twenty years. I can understand why reasonable people may differ on this, but I do not know *any* responsible geologist now engaged in oil exploration who says there is a finite limit that will inhibit ability to find additional supplies of oil and gas in this country in the next twenty years. I know for a fact that most of the oil fields in the United States have produced only about 30 to 50 percent of the oil in the ground. That is because the economics will not permit the industry to recover the balance. Those of you who are engineers, if you want to, can sink a shaft into sand and mine it. You can bring it all to the surface and separate all the oil out of it; you can get 100 percent recovery. And why don't you do it? Because the value of the product you get is not sufficient to pay the cost of recovery. It is an economic problem, not a physical limitation problem.

At this point I want to drop another thought. I do not hear much conversation anywhere about the heavy oil tar sand occurrences around the world. Many, or perhaps most of you, have heard about some heavy oils that occur here in California, where conventional methods may recover 2 to 5 percent of the oil present in the sand. The largest known deposit in the world is in Canada, up in Athabaska, and as best anyone can tell there are somewhere around seven or eight hundred billion barrels of oil in place there. That compares with 30 billion barrels of what we call reserves in the United States, our inventory. It compares with about 150 to 160 billion barrels in Saudi Arabia. Venezuela has a tar belt equal in size to that in Athabaska; on the island of Madagascar and in West Africa there are very large deposits. Nobody really knows how much of this hydrocarbon material is around. It can be recovered, gasoline can be made out of it; but nobody is going to do it until the value of the product exceeds the cost of getting it out. I repeat, there is no physical limitation on the amount of hydrocarbon materials that can be produced and made available to this country.

Chairman: A number of people in the audience want to know how to get an improved long-term energy policy. Who is it—what groups of people, what categories, what individuals in or out of government, in business, in industry—who are the people whose minds have to be changed and how can it best be done?

Swearingen: I think I see some change in the point of view of the Administration, some slow realization. The Congress has been more reluctant to face up to the problem. This comes back to a question as to why our society is different from those of Japan, Germany, or some other countries around the world. The United States has a pluralistic society, and the basic problem is, Americans have not agreed on a set of priorities. We have not agreed whether national defense is more important than welfare. We have not agreed whether an adequate supply of energy is more important than food stamps. We have differing ideas of what government's role should be.

My own view is that the forces that drive government are different from the forces that drive private industry. The government does not ever have to worry about a P&L, it never has to worry about the bottom line. If it spends more than it takes in, what happens? It prints more money. Moreover, it has a claim on everything you and I own, and it can raise taxes to cover its expenses.

But government is not concerned with efficiency; government is concerned more with pleasing a majority of the voters. What I have experienced in Washington (and I have become fairly well acquainted) is that a great many people there believe they can decide how the country's resources are employed, its money is spent, better than you and I as individuals can decide for ourselves. I do not think that government, any government, is very successful in carrying out projects that private industry can do, and I believe one of the most important things in this country is liberty, and that means freedom of decision.

Chairman: Several people in the audience are concerned about oil company profits. The questions are put somewhat in this form: "How does it happen, if the petroleum companies are hard-pressed to find cash to invest in exploration, that they are reporting so much higher earnings now than in previous years?"

Swearingen: One of the things that can never be escaped is what inflation has done to this economy. Maybe I can touch on this inflation question in this fashion. Suppose you ran a grocery store and you had a shelf full of goods, and you knew that the cost of the next shipment of soup was going to be ten percent higher than what you paid for the soup on the shelf; what would you do? Would you sell what was on the shelf at the old price, or would you sell it at a price sufficient to pay for the new shipment? If you sold at the old price, and that happened to everything you had on the shelf and continued long enough, you would end up liquidating your business.

What has happened to us in the oil business is that we are using this rubber yardstick. A dollar today will not buy what a dollar bought ten years ago. And yet oil companies have on the books facilities that were bought five, ten, twenty years ago at dollar-values that existed in those times. Look at the earnings of the industry on a replacement cost basis, and you will find that most of the companies are earning somewhere between a 2 and 5 percent return, and some much lower than that. What is being sold today is the oil in the inventory—I mean underground, not in tanks somewhere, but still in the earth. It will not come out any faster than nature will let it come. The industry is liquidating that inventory at prices for which it cannot be replaced, and if this continues, the oil companies will end up going out of business.

Chairman: On the same subject of profits, a questioner notes that oil companies have a responsibility to their stockholders to seek the highest return possible on investment, and asks, "Is it really true that the way oil company managements can get the highest return on investment is to plow back 90 percent of their earnings (and on occasion perhaps even more than they earn) into exploration?"

Swearingen: I can only speak for myself. I spent forty years in the oil business, I know that business, and I would rather plow back the money our company has into the oil business as I know it, than invest it in aerospace or some other business.

And may I just add one more thing in connection with these profit figures that have been much in the headlines. Very little mention has been made of the fact that most of these profits were earned outside the United States. They were earned by increases in the value of inventories carried abroad, in transit in ships, in tankage, and that kind of thing. They came about as a result of the currency translation figures that are now a part of the SEC's regulation on reporting of earnings. A year ago the dollar was declining and, as a consequence, profits earned in yen, francs, or marks represented fewer dollars than they did this year when the United States dollar was strengthening.

A third point is that profits abroad have been due to a strengthening of product prices resulting from a shortage of supply where foreign governments have, by and large, acted in a way more reflective of the cost circumstances than has our government, which has continued a very onerous system of price control in this country. Look at the profits earned in the United States. I have not seen all of them totalled up yet, but I would be surprised if on the aggregate they were, at the maximum, higher than 10 percent above a year ago. Our own company's U.S. profits were up 8.9 percent in the first quarter.

Chairman: You mentioned government taxes on the consumer's purchase of gasoline. Because those taxes are used for schools, highways and so on, and those functions would still have to be covered if the taxes were removed, would anything beneficial really result?

Swearingen: Yes. The consumer price index would go down [because of the reduction in] the price of gasoline.

Chairman: Question: "When you ask for decontrol to lessen our dependence on imports, are you not really asking for permanent U.S. government support of OPEC's monopoly pricing strategy?"

Swearingen: It is important to realize that buying oil from the OPEC countries today is cheaper than the alternatives that the United States has for supply. The OPEC countries are not setting a price higher than our alternatives. They are setting the price lower. This argument that OPEC is setting the price we in this country have to pay is a completely specious one; it makes political sense, it makes no economic sense.

Chairman: Question: "What do you think will be, roughly, the finding cost for a new barrel of oil in the United States over the next ten years?"

Swearingen: A reasonable estimate of the finding cost of oil today is in the range of $5 a barrel. The trouble with an estimate is that cost and value in the oil-finding business are not the same thing. If you build a house and it burns down the next day, you can build a counterpart of it and it should cost you the same amount of money. In the oil business, a company digs ten wells and it may find nothing or it may find a huge oil field. Cost and value are not the same thing; they differ in time, they differ with the skills, they differ with the luck of the people in the business. So when I talk to you about a $5 figure, this is a typical figure; I cannot guarantee that it is an average one.

Now then, what will it be over the next ten years? I don't know, but let me start with this: if inflation proceeds at the rate of only seven percent a year, the value of money in ten years will be half what it is today. In other words, the cost in dollars to accomplish the same physical act will be twice, ten years from now, what it is today. And if inflation proceeds at a higher rate, that number will accelerate accordingly. In addition to this, the search for oil is going into more difficult areas and there is a physical factor involved as well. As it is necessary to go deeper and into more remote and difficult areas to find oil, that factor will be added on to anything else that occurs as a result of general inflation.

Chairman: Question: "The space program was government-funded, and virtually everyone involved benefitted because of the large sums of money plowed back into the economy. The public did not have to make any personal sacrifices. However, any decontrol of oil prices and, consequently, higher prices for the public, results in direct impact to the pocketbooks of the public. Since human nature is basically selfish, how do you expect the nation to pull together behind such an energy policy?"

Swearingen: Does anybody in this room, including the person who wrote that question, really believe that the sky rockets we fired off to the moon were free? Did not cost anything? Do you really believe that? Where do you think the money came from? It came out of your pockets and out of my pocket in the form of taxes. This comparison [with the space program] makes no sense to me.

Chairman: In defense of the person who asked that question, he may have meant not that he, but the public in general, thinks that the space program did not cost them anything; and that is part of the problem—the ignorance of the general public.

Swearingen: This is where the leaders of our country are doing the nation a disservice. When they want to berate the oil companies, berate big business in general, they are doing it with an eye focused on the next election in 1980. It is popular in this country to tear down the big for the benefit of the small. And when you tear down the big, the question is, what will replace it? My reading of history has been that most of the revolutions that have occurred over time have not been against portions of the population, but against tyrannical governments. Put the government in the position of running all America's business activity and personal affairs, and this country is into some very big trouble. I hope we never get to that stage.

Selwyn Enzer

Energy Policy: A Failure of Foresight

A tradition in business and government is to regard the long-range future—the time period from about 5 to 20 years—as fundamentally unpredictable and hence beyond meaningful analysis. In this paper I would like to challenge the notion that the long-term future is the exclusive domain of cranks and charlatans. And I would like to do so by taking energy as an example.

Let me begin by pointing out that the current energy situation could have been anticipated in the pre-embargo early '70s as one of the many possible alternatives. At that time, many people not only predicted today's energy situation, but also foresaw that energy conditions would continue to deteriorate. Despite the demonstrable accuracy of those predictions, they could, of course, have been wrong. That is, the current energy conditions did not have to occur. Numerous choices were available to the United States in the early '70s which, if implemented, could have altered the present state of affairs. Unfortunately, none of these choices was adopted.

Today's energy situation is the result of a failure of foresight, not because current conditions could not be anticipated, but because the

---

Selwyn Enzer *is Associate Director of the Center for Futures Research in the University of Southern California's Graduate School of Business, where he developed the INTERAX model as a basic tool of applied long-range study and planning. Formerly with the Institute for the Future, he has conducted seminars in futures research for industry and government in the United States, Canada, Europe and Japan.*

United States did not act upon available foresight five years ago. This failure of foresight must not be repeated. While energy conditions in the future will undergo many changes, they need not be calamitous. There are numerous choices available to the United States now which can alter the energy picture in the 1980s and beyond. Unfortunately, these energy choices all have some negative impacts on near-term conditions and only in the longer term are their benefits fully realized. But, before suggesting a model for analyzing these choices, let us look back a few years to see what was anticipated about the present.

The Early '70s

Prior to the oil developments of 1973 there were many forecasts indicating that U.S. dependence on foreign oil was growing at an alarming rate. A compendium of pre-1973 U.S. energy forecasts is shown in Figure 1. There was little variation among the nearly twenty

Fig. 1. Pre-Crisis U.S. Energy Outlook—Composite Forecasts

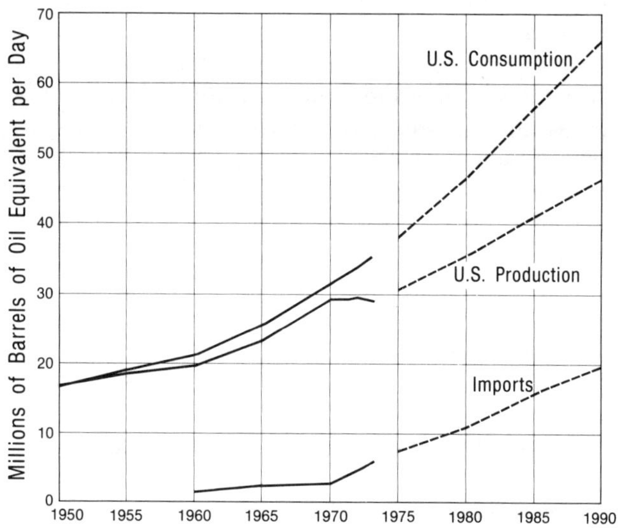

Source: Selwyn Enzer, "Some Societal Impacts of Alternative Energy Policies," Report WP-21, Institute for the Future, Menlo Park, CA, June 1975.

public and private organizations whose forecasts were used to prepare this composite. In short, before the embargo there was general agreement about the rate at which U.S. energy imports would grow in the future.

This generally gloomy consensus among energy experts was influenced by the fact that nuclear energy was coming along more slowly than had been expected, and it appeared unlikely that this trend would change. Many utilities had developed contingency plans for allocating energy—plans that they felt would be needed in the early 1980s. They recognized that even if there were to be a sudden breakthrough in nuclear energy (or any other alternative) it would still take at least fifteen years to make sizeable amounts of new energy available. This would be too long a delay to avoid at least a temporary crunch, given the rate of growth of energy consumption in the United States and the slowdown in domestic oil and natural gas production. The experts saw all this in the early 1970s, yet it was impossible to alert a sleeping America to the potential criticalness of the emerging conditions.

In 1973, when President Nixon declared that the country faced a serious energy crisis in the future, many Americans thought his statement was a political ploy designed to take the focus away from Watergate. With his credibility eroding rapidly, Nixon could do very little to mobilize the forces needed to change the course of the energy situation.

After the OPEC oil embargo, the general perception of long-term energy conditions seemed to change, at least temporarily. It was thought that the OPEC actions would force the United States to initiate programs to increase energy production and conservation. This can be seen in the 1974 National Academy of Engineering study of U.S. energy prospects, which concluded:

> Projections of domestic supply and demand made before the fall of 1973 predicted the need for large and increasing imports of oil and gas with attendant problems of potential supply interruptions, higher costs, increasing effects on balance of payments, and questions of worldwide productive capacity for oil. The oil import embargo in late 1973 and the concurrent sharp increase in oil prices by essentially all of the oil-exporting nations have produced a substantial change in U.S. and worldwide thinking and lead to further consideration of means for reducing energy demands in the United States and increasing domestic energy production.
>
> While domestic energy production cannot be increased abruptly, in time it can be expanded over the present levels or those previously projected for 1985. Production levels of about 49 MBPD by 1985 seem feasible in contrast to projections of 40 MBPD made in mid-1973 and actual 1973 levels of about 30 MBPD. However, the costs will be high,

> and such a program will strain U.S. resources and will place considerable stress on the environment.*

The Project Independence team formed by President Nixon sounded a similarly optimistic note.** It concluded that the energy crisis could be avoided over the next two decades with a program that emphasized higher oil prices, accelerated production, and conservation.

The post-embargo forecasts by the National Academy of Engineering and Project Independence are illustrated in Figures 2 and 3. In retrospect

Fig. 2. Post-Crisis U.S. Energy Outlook
(National Academy of Engineering, 1974)

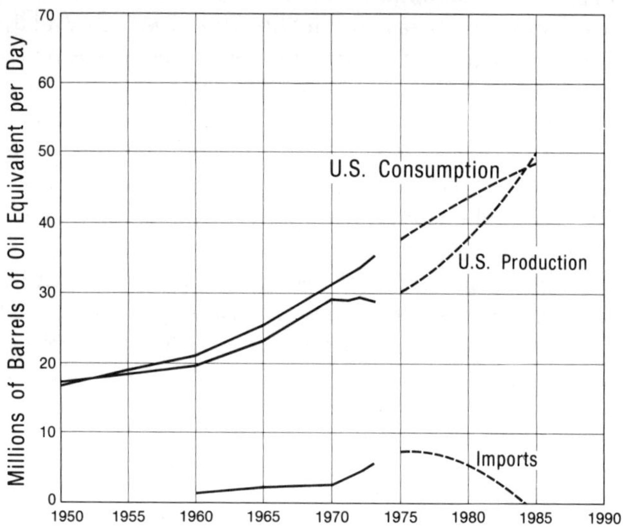

both of these projections appear highly optimistic. The former overestimates the power of technology, and the latter the power of the price mechanism. However, both projections suggest a sharp improvement in outlook for the U.S. energy situation. But, as the United States moved from 1974 to 1975 to 1976, it became apparent that these expectations were based on false hopes. Figure 4 graphically illustrates the disap-

*U.S. Energy Prospects—An Engineering Viewpoint," National Academy of Engineering, Washington, D.C., 1974.

**Energy Perspectives," U.S. Department of the Interior, February 1975.

Fig. 3. Post-Crisis U.S. Energy Outlook
(Project Independence, Feb. 1975)

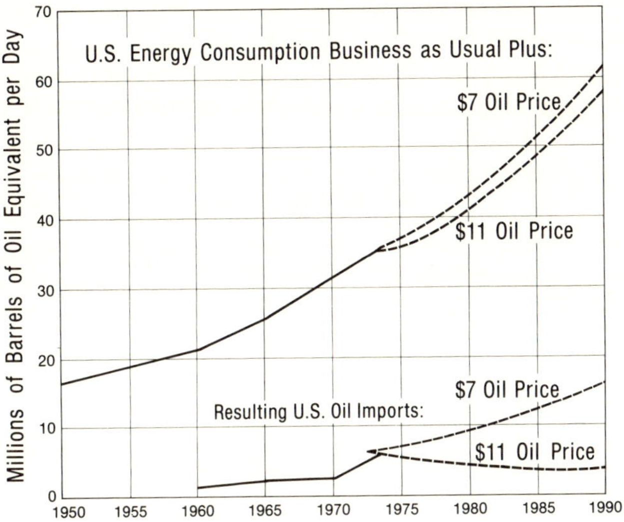

Fig. 4. Current U.S. Energy Consumption

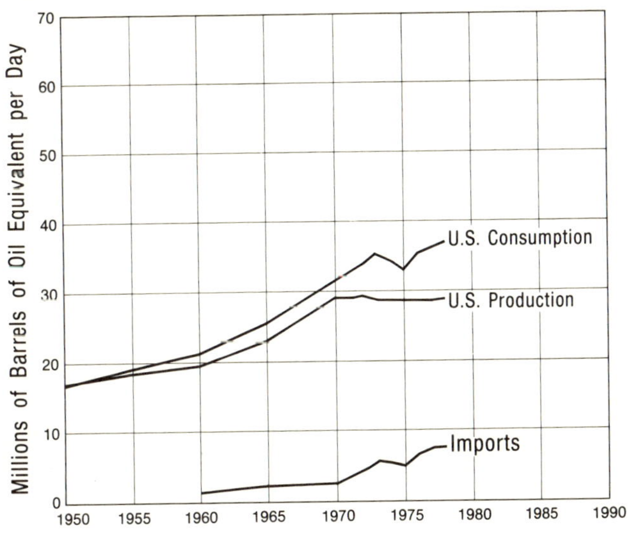

pointing performance of U.S. policy in energy production, consumption, and imports over the last six years.

This is not just another forecast gone wrong. This is a failure of government to respond to a critical long-term issue. And it is a failure on the part of the expert community to anticipate the actual response to energy forecasts, thereby raising false expectations about the long-range future.

What Went Wrong

Incorrect forecasts are the result of two possible errors: either surprising changes occurred, or the system did not behave in the manner anticipated by the forecaster. Surprises come from the occurrence of developments that were not considered likely, from the nonccurrence of developments that were expected, and from developments that were never thought of in the first place.

Based on 1974 expectations, some surprises have occurred in the last five years. Most of these, however, were positive developments. The natural gas shortage proved to be less critical than originally anticipated. Mexican oil deposits proved to be much greater than anticipated. North Sea oil and Alaska oil came along on schedule. By current low standards of expectations, these developments must be regarded as at least modestly surprising. On the other hand, the developments in Iran and the nuclear accident at Three Mile Island were surprises to almost everyone. Perhaps the biggest surprise in the past five years is that domestic energy production not only failed to increase over the low pre-embargo projections, but actually fell substantially below that level.

The 1974 revisions in long-range energy forecasts were based primarily on the anticipation that the United States would act in an "economically prudent" manner, that is, take steps to reduce domestic consumption and to increase domestic production, while at the same time seeking alternative energy sources for very long-term needs. Without intending any overstatement, I think that the national performance in this regard can at best be described as dismal.

Rather than focusing on means to expand domestic energy production, pursue new energy sources, and promote conservation, U.S. regulatory policy addressed near-term energy prices and profits. This did little more than create an accounting nightmare in the energy industry in which oil took on demographic characteristics—young, old, and adolescent. But one cannot realistically claim that options for altering energy trends did not exist. There were many alternatives, and they were widely known.* They included:

- Intensified search for new domestic reserves
- Greater use of plentiful fuels
- Development of new energy sources
- Conservation programs (e.g., mass transportation, and more efficient systems).

These options could have been pursued with a variety of policies that used such tools as price mechanisms, non-price allocation schemes, regulations, subsidies, tax incentives, direct government investment (e.g., in mass transportation), and combinations of these. These approaches all have strengths and weaknesses. None of them is easy and none is without hardships.

This point requires some elaboration. The nub of the energy crisis was (and still is) America's growing dependence on foreign oil, the flow of which, if interrupted for even a brief period, causes chaos—a glimpse of which was initially seen in California. To reverse this growing dependence, America will have to produce more oil and consume less, and eventually switch away from oil to other energy sources. To many Americans, reduced oil consumption translates directly into less mobility and a reduced standard of living. The democratic process has great difficulty in allocating disbenefits such as these. Hence, in 1974, it was easier to discount the future, addressing only near-term issues and counting on the tried-and-true policy of "muddling through" to take care of the longer term.

Today some new options exist that did not exist then. Some options that were considered viable then may no longer be feasible today. The time available for change has been reduced by five years. But perhaps the most critical change is that today credibility among all interest groups has been severely eroded. (Trust among the energy companies, the federal government, state governments, environmentalists, and the public, is at all times lows.) But the choices America faces today are similar to those faced in 1974. If America was not able to adopt the stern measures demanded by a farsighted posture in 1974, what can make their adoption more likely now?

---

*In addition to the previous references see, for example, Edward Teller, "Energy—A Plan for Action," Commission on Critical Choices for Americans, New York, New York, April 1975; and James O'Toole, "Energy and Social Change," M.I.T. Press, Cambridge, Mass., 1976.

Urgently Needed—A Farsighted Posture

The experience of the 1970s has made it quite clear that long-term energy conditions will not right themselves without major policy interventions. There are no forces in the world economic system that will bring all energy supply and demand into balance painlessly. What is needed is a commitment on the part of the United States to regaining control of its energy situation over the long term. This commitment can be supported by an even-handed analysis of long-range alternatives, including an understanding of the full range of consequences that might flow from these alternatives. With the proper commitment and a realistic appraisal of the long-term consequences, America may be able to develop the consensus needed to adopt an energy policy that properly balances near-term and longer-term objectives.

The experience of recent events notwithstanding, false hope springs eternal in the breast of the re-election minded politician. So it would not be surprising to find politicians voicing hopes that the current round of oil price increases will produce the incentive needed to bring long-term energy supply and demand into balance in the United States. While there is no doubt that a *real* price increase in energy would promote this balance, these recent increases will trigger the next wave of inflation and cost-of-living adjustments. These economic reactions will reduce the impact of higher prices in promoting conservation and intensifying the search for additional energy. In other words, price mechanisms cannot be relied upon to work their magic unless indexing measures which are built into the system, and which stabilize existing consumption and production patterns, are constrained.

However, politicians are likely to accept the notion that the current price rises will cure the long-term energy problem because it is the easiest solution for their constituents to accept. These false hopes only interfere with developing and carrying out an effective plan for America's energy future—which requires a significant improvement in our farsightedness. There are methods and procedures that can assist government and industry in achieving a more farsighted posture, but effective use of these methods requires a basic change in the way institutions explore alternative futures. Helping institutions make this change has been one of the goals of futures research since its inception.

A Futures Research Approach

Futures research came into being in the early 1960s largely because long-range forecasts using conventional methods usually proved to be inaccurate. Futures researchers reasoned that these forecasts were prone to failure because they were based exclusively on the past. While historical models may suggest the nominal future course of a trend like energy consumption, they fail to show how future political, social, economic or technical developments could alter this course or how its direction could be changed by policy options. The methods of futures research are designed to rectify these omissions by augmenting conventional forecasts with judgmental insights about the future itself. And because issues as broad as energy can be affected by a wide spectrum of developments, the judgmental insights must be multidisciplinary, including changes in economics, government regulations, social conditions, technology, demography, etc. Basically, futurists consider it more important to "study" the future than to "forecast" it. This is based on the observable fact that many futures are possible and that the alternative that ultimately comes to pass depends upon the outcome of foreseeable, but inherently uncertain events, and on policy choices.

The Center for Futures Research has developed a procedure for evaluating long-term strategic issues that accommodates all of these factors. This procedure, called INTERAX, uses probabilistic computer models to generate plausible individual scenarios on a year-by-year basis. Users interact with the scenarios as they are being generated by introducing new policies into the models at the beginning of any year.

INTERAX captures the three major components that shape the future in the manner shown in Figure 5.

Fig. 5. Energy Version of INTERAX

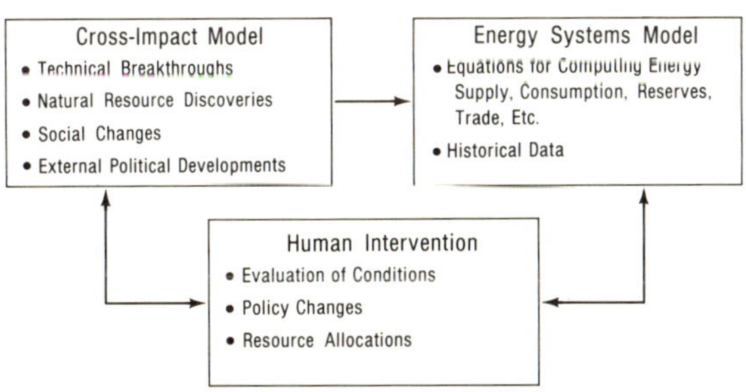

The Energy Systems Model represents current data and trends in energy production and consumption for the United States and the world. At the beginning of each year this model would project the levels of consumption, production, inventory, trade, etc. These projections could be obtained from a wide variety of sources in the form of econometric or system dynamics models.

The Cross-Impact Model accounts for the occurrence of foreseeable events that could affect the future course of the energy system. These events may range from technological breakthroughs to wars and resource discoveries. Because these events are interrelated, in fact and in the model, the occurrence of any one of them not only affects the energy systems model but also affects the subsequent probabilities of the other events.

In the Human Intervention component, users simulate various interest groups—government, energy companies, environmentalists, etc.—who act to shape energy futures. Within the control of the simulation group is a host of energy policy choices. Prior to each year of the simulation, the interactors evaluate the scenario being created and then formulate a policy they believe would be appropriate for the interest group they are representing. One can imagine how the realism of the process is increased by giving users the opportunity to change policies in each projected future period. The content and impact of real policy deliberations simply cannot be captured by merely running a computer model with predetermined strategies. Although it is easy to prè-specify certain contingent responses, it is obviously impossible to anticipate and prespecify a response to every situation that can occur. This is the essence of interactive modeling: people periodically adjust the policies that guide a forecasting model on a year-by-year basis to create an individual scenario in a manner that simulates real world processes.

How to Handle Uncertainty

The open-ended nature of this procedure conflicts with conventional approaches to planning and forecasting in several ways. The most important of these is in the handling of uncertainty. Planning, as presently practiced in government and industry, is designed to reduce uncertainties that may otherwise occur in the future which may block the fulfillment of a goal. Unfortunately, when conventional planning procedures are used in long-range forecasting, they produce results that often reflect much greater certainty than exists in the real world. In such cases, surprise becomes the rule rather than the exception. The INTERAX procedure emphasizes capturing the uncertainty that exists in the real world, and it evaluates alternative scenarios in the presence of that uncertainty.

Better handling of uncertainty is the key to developing a farsighted posture. This is the essential difference between *foresight* (the predictive activity experts engage in when they indicate the most likely future) and *farsight* (the understanding of how alternative conditions might develop in the future and how these conditions can be changed). In the energy area there is too much effort toward achieving foresight and too little in developing farsight. The nuclear accident at Three Mile Island is a good example of how overconfidence, based on too much certainty (foresight), can block the development of a farsighted posture. Not only was the public told that the risk of a serious accident at the nuclear plant was miniscule, but because nuclear safety institutions were officially committed to that belief, no real emergency procedures were developed. Neither the General Public Utilities personnel nor the experts at NRC knew what to do during the critical days following the failure of the reactor's cooling systems. Why? Because they had not really planned for such a possibility. When faced with the prospect of evacuating the area around the crippled plant it became apparent that no plan existed. Why? Because they did not think such a situation could realistically occur.

It is difficult to imagine that the public would ever have supported a nuclear program so hazardous that "real" mass evacuation procedures have to be developed to cope with the threat. Would you support the development of a nuclear plant in your neighborhood after the Civilian Defense authorities visited your home to make sure you and your family were thoroughly familiar with the evacuation procedures to be used in the event of a nuclear disaster—which, by the way, is virtually impossible in the first place?

I suspect that most people will reject further development of nuclear power plants under these conditions—in which case, the Three Mile Island accident will prove to be a pivotal event in changing the trend in U.S. energy production. Significantly, this contingency would not be detected, even in a future-oriented analysis, if the probability of such an accident were perceived to be so small as to be negligible.

It is therefore highly important that uncertainties be realistically included in long-range forecasting and planning. INTERAX is one way to capture this uncertainty. Human interactors attempting to use this system to study alternative energy futures would face an analytic situation that contains the uncertainty they would find in the real world. The results obtained from this process are not pat answers. Rather, from our experiences using this approach to study world food futures, we find that planners come away with a greater appreciation of alternative possible outcomes, policy choices, and follow-up actions that may be necessary in the face of subsequent surprises, some of which will invariably occur.*

The primary benefit obtained from this procedure is that it forces explicit evaluation of long-range outcomes on a comprehensive basis. Because the procedure analyzes the full range of impacts that follow from an event or policy change, it prevents piece-meal thinking. With this procedure an analyst has to consider the undesirable as well as the intended consequences and timing of policy changes on subsequent conditions. The whole idea is to make the images of the future so clear that the consequences of alternative policies become obvious in prospect.

A Future-Oriented Outlook

A future-oriented assessment of the energy situation, conducted through an approach like INTERAX, would illuminate three different phases:

- The next ten to twenty years—a period in which U.S. consumption and production of oil must be brought into reasonable balance;
- The period from twenty to seventy-five years out—a period in which more plentiful energy sources must be developed to replace diminishing oil and natural gas supplies;
- The period beyond the next seventy-five years, during which new energy sources must be made available.

A farsighted posture requires a strategy that responds to the needs while addressing the alternatives available in each of these time periods.

One might logically question whether the current U.S. energy strategy addresses each of these phases of the energy situation. A partial answer to this question can be seen in an energy forecast recently prepared by Exxon.** The Exxon forecast, presented in Figure 6, addresses the period to 1990, and it embodies most of the recommendations presented in this conference as they affect conditions in this time period. It assumes that: 1) overall energy consumption in 1990 will be 25 percent less than the

*S. Enzer, R. Drobnick, and S. Alter, "Neither Feast Nor Famine," Chapter 5, Lexington Books, Lexington, Mass., 1978.

**U.S. Energy Outlook—1979-1990," Exxon Company, December 1978.

Fig. 6. U.S. Energy Outlook, 1979-1990 *(Exxon Company, Dec. 1978)*

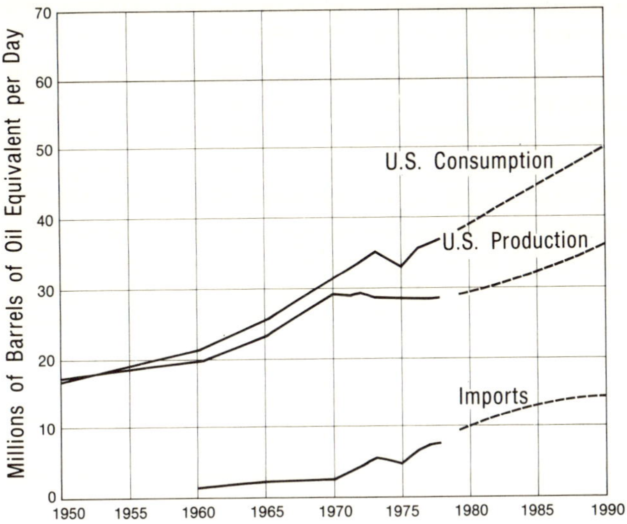

pre-crisis projection, primarily as a result of price-induced conservation; 2) domestic coal production will double between 1978 and 1990; 3) nuclear energy production will nearly quadruple in this time period; 4) synthetics will be commercially available after 1984; and 5) imports—mainly OPEC oil—will provide for the shortfall, about 30 percent of U.S. energy consumption.

The Exxon forecast includes the production increases likely to be realized from the enactment of most of the programs recommended by the various speakers in this colloquium. Yet, without even considering the impacts of a curtailment in nuclear development in this time-period, any even-handed assessment of this forecast would conclude that the future it describes is unacceptable. The United States cannot afford to be in a position where, by the end of the next decade, it derives 41 percent of its energy from oil when it has to import 64 percent of that oil, especially when such imports rely on a system as volatile and fragile as one managed by OPEC. The experience of the 1970s offers sufficient evidence to prove this point.

But how can this future be changed? Because it is doubtful that programs to increase domestic energy production over the next decade will reduce U.S. oil dependency to an acceptable level, then U.S. oil consumption must be reduced. A program to reduce oil consumption will not necessarily reduce energy consumption. It would, of course, seek to conserve energy wherever possible, but its primary thrust would be to convert oil-consuming activities to the use of energy forms that are domestically more plentiful, even though these may be rapidly running out also.

A program to reduce U.S. oil consumption should have two components—conversion from oil to energy forms that are domestically more plentiful, and conservation in those areas where conversion is not feasible. Some of the components of a program to reduce oil consumption are relatively easy to accomplish. These include switching from oil in stationary heating applications to natural gas, or coal, wherever possible. The potential oil savings from such conversion are enormous. Electrical power plants and space heating alone represent about one-third of all U.S. oil consumption. These applications are relatively easy to convert to other forms. Methanol in space heating can save more oil than its highly publicized use in automobiles. Another target of oil conservation is the transportation sector which takes about 57 percent of all U.S. oil consumption. What is needed here are systems that are more efficient than those currently in use. Examples include mass transportation designed expressly to reduce energy consumption (as opposed to providing universal access to transportation), greater use of railroads for intercity shipping (trucks-on-train), etc.

These are but a few aspects of what a comprehensive program to reduce oil consumption must explore in detail. Such a program would ease oil consumption while further burdening domestic sources of energy. What is lacking is the recognition that a program to reduce U.S. oil consumption is essential to national (and perhaps even international) economic stability over the next two decades. During this period of stability the U.S. would be able to devote appropriate resources to developing solutions for the longer term. But only by proceeding with a program to reduce oil consumption in the near term can the United States regain control of its energy future in the longer term.

Reducing oil consumption in the near term will not be easy or pleasant, but if U.S. oil imports continue to grow at their expected rates, supply disruptions and shortages will become the rule, not the exception. The consequences of such conditions can provoke military as well as economic crises. The issue is not whether to reduce oil consumption, but how to do so. To some extent this can be accomplished through higher prices, but price alone will not always produce timely or socially desirable changes. Oil reduction programs must be initiated even if such programs are not currently economically viable or as attractive as their oil intensive counterparts.

America must provide alternatives for reducing its oil consumption before strangling on shortages of supply. The tragedy of the current forecasts is that they clearly indicate that America has learned very little from the oil experiences of the 1970s.

Conclusion

I began this discussion by challenging the notion that the future is beyond meaningful analysis. I hope this brief reflection on the '70s convincingly shows that while the future cannot be predicted precisely, it can be better understood and more effectively controlled. However, it is not sufficient merely to create forecasts that indicate the need for political action but fail to motivate such action. Indeed, there is no guarantee that a future-oriented analysis such as INTERAX or any other technique can create this commitment. All such analyses can do is make future considerations a more important factor in current policy choices.

The key to the U.S. energy situation over the longer term may well depend on the degree to which the nation reduces dependence on foreign oil in the next decade or so. This reduced dependence can come about only from increased domestic production and decreased consumption. Since the prospects for sizeable increases in production are small, current policy must focus on reducing consumption. However, any policy that reduces oil consumption will cause an immediate reduction in the U.S. standard of living. Such policies are never politically attractive, and if they can be delayed, the political process will avoid them.

Experience has shown that the enemies of tough political action are false hopes. False hopes can be attacked with farsight. A future-oriented analytic approach can be highly effective in achieving this posture because it distinguishes false hopes from viable courses of action. Such farsight is essential to support the policies needed to assure that U.S. energy futures are realistic and consistent with American interests and objectives.

Discussion

From the floor: INTERAX, of course, is a new technique, and it cannot yet show its value in the energy situation. Have you tried it on anything else, such as the stock market?

Enzer: Yes, this technique was developed in the context of transportation planning for the State of California and, more recently, world food. It is currently being used in world trade and also as a planning tool for about ten large corporations.

I do not want to overstate its potential. It definitely is a research tool; there is much that is not known about this process. It is an open-ended problem-solving process; that is, it does not yield one answer, it yields many answers. It is used to explore multi-networks of alternatives.

Open-ended processes are not really understood. They have never been studied seriously. The approach has always been to come up with a definite problem that had an optimal answer and solve for that optimal answer. If the United States had precise goals for the energy situation in 1990 or 2000 which it was agreed had to be met, this would be a very easy situation to attack, to resolve. But there is no consensus on goals. We are wrestling with alternatives. INTERAX can help, but it is in the research stage. We have to learn from it.

From the floor: Many of the points you made are well taken—about trying to deal with the problems that are not included in either the models or the formal analysis but are going to be very important in affecting the future—but it is not quite true that this has not been done in the past. If one looks at the National Petroleum Council's study or the National Academy of Engineering study, or even the Project Independence numbers you have talked about, one finds they said that if the United States were to do certain things, like decontrol oil prices, increase the oil leasing rights, and so forth, along with taking some conservation options, then maybe by 1985 it could get to a desired specific objective. If steps were not taken, it would not reach that objective.

So the difficulty is not that the tools and the analysis are weak or are not helping people think through these problems. Rather, the difficulty is in trying to incorporate all the various assumptions that you were talking about. And I do not see how your method will overcome the fact that people always walk away with one idea, no matter how complicated the study might have been.

Enzer: I did not mean to imply that the Project Independence forecasts or the National Academy of Engineering forecasts were incorrect because their models were basically wrong. Their models operated on a set of assumptions that unfortunately had to be static because they were deterministic models. Hence, certain policy assumptions were used in the models and the solutions were ground out.

The Project Independence forecast I showed (Figure 3) was the business-as-usual forecast with the difference being price. It did not include intensified search for a new supply of oil, nor did it include any emphasis on the conservation program, but it did have the factors that you identify—namely passing rewards through to the producers and trying to increase production incentives that way.

What is apparent is that these models did not include any of the social factors, or other possible sources of change—dynamic things that could affect the projections once the assumptions were locked into the model. One must have a plan to follow for the system, but it has to be a very dynamic plan. You must know what your options are at all times, just as when you play chess and are employing a certain strategy: you may think you know where you are going and what you are trying to do, but if unexpected things occur, then you must be ready to back off from that position and take a different course. That is what is not captured by existing models, and is one of the major contributions that Futures Research has made in this area. I think in the course of time people will learn how to develop and use dynamic models effectively; I am not sure it can be done very well yet.

From the floor: I am interested in your use of the National Academy of Engineering study since I happened to be chairman of the committee that prepared it. We started that in 1973, before the embargo and the price increases. It was not intended to be a forecast. (I rather object to the use of that word in connection with the study.) What we tried to do was to point out what could be accomplished with, not a long-range plan, but a rather short, very important plan that included getting on with synthetic fuels. I might add, a number of suggestions we made are not greatly dissimilar from what has been said here these past days.

But the fact is, nobody really paid any attention. I do not know that the results we projected would have come about if somebody had, but current conditions would certainly be a lot closer to what we projected. So my question now is, how can Futures Research help to get the country to do what it ought to do instead of the same old exercise, so that six years from now we or other people will be having this same discussion all over again?

Enzer: Clearly the National Academy of Engineering did not make a forecast of what would happen in any prophetic way; what they were trying to show was the power of technology to respond to this problem, and they presented their statement in that regard. It may have been an overstatement of the capacity of technology as it was available, but that is an engineer's prerogative. The point I was trying to make—and that is why I read from their concluding statement—is that in a political environment, such forecasts do not have the potential for relating back to decisions and decision-making processes in a holistic way. Rather, they create false hopes in a politician who is looking for a false hope. "If I can defer taking action today, I have this invisible, muddling-through, magic process that will take care of the long term. Right now I have to be re-elected." I think that is one of the issues, and one of the traps we fell into.

From the floor: I should think that perhaps your model is a communications device.

Enzer: It is indeed to a large extent a communication device. But it is also a learning device by which those people who are interacting with it in fact learn from it. The objectives or goals that a person sets for himself do not come about except through resources that he controls or through changes that come to pass. One thing that becomes quite apparent is that the consequence of inaction is a continuation of present trends.

One really needs a little more exposure to this kind of process to appreciate it, but there is another interesting thing it brings out that I would like to highlight. Suppose you are following one of the possible paths in that big network of alternative scenarios that can evolve over time, and you get to a point perhaps ten years down the road. You then try to describe the situation as you think it would appear in the future, and you describe what it means to you. I would argue that you would regard this situation as one which, if it were to occur, would have been obvious to you in retrospect. That is, looking back and using your 20-20 hindsight it would be an obvious path. All that means is that you could have anticipated it in foresight. So this is really a method of forcing you to confront your own holistic view of an emerging situation and try to make very clear the kinds of things that could be obvious to you in prospect, including the kinds of traps you could fall into and where they might come from.

All this will not help much with a Three Mile Island sort of event, because if the probability is set so low that in essence it cannot happen, then it affects nothing. In that case, there is too much certainty about what will happen; and too much certainty about what will happen drives out any farsightedness—which involves understanding what might happen and what can be done about the various alternatives.

GLOSSARY

AEC — Atomic Energy Commission
Btu — British thermal unit
CIA — Central Intelligence Agency
CPI — Consumer Price Index
DOE — Department of Energy
DRI — Data Resources, Inc.
EEC — European Economic Community ("Common Market")
EOR — enhanced oil recovery
ERC — Energy Regulatory Commission
ERDA — Energy Research and Development Administration
FERC — Federal Energy Regulatory Commission
FUA — Fuel Use Act
GDP — gross domestic product
GNP — gross national product
GWe — gigawatts electric (one billion watts)
IAEA — International Atomic Energy Agency
LNG — liquefied natural gas
MBPD — million barrels per day
MMBOE — million barrels of oil equivalent
MOPPS — Market Oriented Program Planning Study
Mtoe — million tons of oil equivalent
Mw — megawatt (million watts)
NEA — National Energy Act
NEP — National Energy Plan
NRC — Nuclear Regulatory Commission
OAPEC — Organization of Arabian Petroleum Exporting Countries: Saudi Arabia, Kuwait, Qatar, the United Arab Emirates, Iraq, Libya, Algeria, Egypt, Syria, Bahrain
OECD — Organization for Economic Cooperation and Development
OPEC — Organization of Petroleum Exporting Countries: Saudi Arabia, Kuwait, Qatar, United Arab Emirates, Iraq, Libya, Algeria, Iran, Indonesia, Nigeria, Gabon, Venezuela, Ecuador
OTEC — ocean thermal energy conversion, a technology for producing electricity from energy stored in tropical ocean waters
quad — quadrillion Btu's
R&D — research and development
SNG — substitute natural gas